Defense Manufacturing in 2010 and Beyond

MEETING THE CHANGING NEEDS OF NATIONAL DEFENSE

Committee on Defense Manufacturing in 2010 and Beyond

Board on Manufacturing and Engineering Design

Commission on Engineering and Technical Systems

National Research Council

NATIONAL ACADEMY PRESS
Washington, D.C. 1999

NOTICE: The project that is the subject of this report was approved by the Governing Board of the National Research Council, whose members are drawn from the councils of the National Academy of Sciences, the National Academy of Engineering, and the Institute of Medicine. The members of the committee responsible for the report were chosen for their special competencies and with regard for appropriate balance.

The National Academy of Sciences is a private, nonprofit, self-perpetuating society of distinguished scholars engaged in scientific and engineering research, dedicated to the furtherance of science and technology and to their use for the general welfare. Upon the authority of the charter granted to it by the Congress in 1863, the Academy has a mandate that requires it to advise the federal government on scientific and technical matters. Dr. Bruce Alberts is president of the National Academy of Sciences.

The National Academy of Engineering was established in 1964, under the charter of the National Academy of Sciences, as a parallel organization of outstanding engineers. It is autonomous in its administration and in the selection of its members, sharing with the National Academy of Sciences the responsibility for advising the federal government. The National Academy of Engineering also sponsors engineering programs aimed at meeting national needs, encourages education and research, and recognizes the superior achievements of engineers. Dr. William Wulf is president of the National Academy of Engineering.

The Institute of Medicine was established in 1970 by the National Academy of Sciences to secure the services of eminent members of appropriate professions in the examination of policy matters pertaining to the health of the public. The Institute acts under the responsibility given to the National Academy of Sciences by its congressional charter to be an advisor to the federal government and, upon its own initiative, to identify issues of medical care, research, and education. Dr. Kenneth I. Shine is president of the Institute of Medicine.

The National Research Council was organized by the National Academy of Sciences in 1916 to associate the broad community of science and technology with the Academy's purposes of furthering knowledge and advising the federal government. Functioning in accordance with general policies determined by the Academy, the Council has become the principal operating agency of both the National Academy of Sciences and the National Academy of Engineering in providing services to the government, the public, and the scientific and engineering communities. The Council is administered jointly by both Academies and the Institute of Medicine. Dr. Bruce Alberts and Dr. William Wulf are chair and vice chair, respectively, of the National Research Council.

This study by the Board on Manufacturing and Engineering Design was conducted under contract no. N00014-96-D-0301 (Task Order 02) with the U.S. Air Force, the U.S. Army, the U.S. Navy, and the Defense Logistics Agency. Any opinions, findings, conclusions, or recommendations expressed in this publication are those of the authors and do not necessarily reflect the view of the organizations or agencies that provided support for the project.

Library of Congress Catalog Card Number 99-60164
International Standard Book Number 0-309-06376-0

Available in limited supply from:
Board on Manufacturing and Engineering Design
2101 Constitution Avenue, NW
Washington, D.C. 20418
202-334-3505
email: bmaed@nas.edu

Additional copies are available for sale from:
National Academy Press
Box 285
2101 Constitution Avenue, N.W.
Washington, D.C. 20055
800-624-6242
202-334-3313 (in the Washington, D.C. metropolitan area)
http://www.nap.edu

Copyright 1999 by the National Academy of Sciences. All rights reserved.
Printed in the United States of America.

COMMITTEE ON DEFENSE MANUFACTURING IN 2010 AND BEYOND

ALTON D. SLAY (chair), Slay Enterprises, Inc., Warrenton, Virginia
HENRY ALBERTS, Defense Systems Management College, Fort Belvoir, Virginia
ROBERT F. BESCHER, Pratt & Whitney, West Palm Beach, Florida
WILLIAM GIBBS, Electric Boat Corporation, Groton, Connecticut
WESLEY L. HARRIS, Massachusetts Institute of Technology, Cambridge
DAVID LANDO, Lucent Technologies, Murray Hill, New Jersey
ARIS MELISSARATOS, Westinghouse Electric Company, Pittsburgh, Pennsylvania
FREDERICK J. MICHEL, NGM Knowledge Systems, Alexandria, Virginia
J. DAVID MITCHELL, Rockwell International (retired), Grand Junction, Colorado
DEBORAH S. NIGHTINGALE, Massachusetts Institute of Technology, Cambridge
DEAN RHOADS, Merrill Lynch, New York, New York
RICHARD SEUBERT, The Boeing Company, Seattle, Washington

Board on Manufacturing and Engineering Design Staff

ROBERT M. RUSNAK, Senior Program Officer
BONNIE A. SCARBOROUGH, Program Officer
THOMAS E. MUNNS, Associate Director
AIDA C. NEEL, Senior Project Assistant
CHARLES HACH, Program Officer
LOIS LOBO, Research Associate

Board on Manufacturing and Engineering Design Liaison

DOROTHY COMASSAR, GE Aircraft Engines, Cincinnati, Ohio

Government Liaisons

LEO PLONSKY, Office of Naval Research, Philadelphia, Pennsylvania
GERALD SHUMAKER, Wright-Patterson AFB, Ohio
CHARLES E. STUART, U.S. Department of Energy, Washington, D.C.

BOARD ON MANUFACTURING AND ENGINEERING DESIGN

F. STAN SETTLES (chair), University of Southern California, Los Angeles
ERNEST R. BLOOD, Caterpillar, Inc., Mossville, Illinois
JOHN BOLLINGER, University of Wisconsin, Madison
JOHN CHIPMAN, University of Minnesota, Minneapolis
DOROTHY COMASSAR, GE Aircraft Engines, Cincinnati, Ohio
ROBERT A. DAVIS, The Boeing Company, Seattle, Washington
GARY L. DENMAN, GRC International, Inc., Vienna, Virginia
ROBERT EAGAN, Sandia National Laboratories, Albuquerque, New Mexico
MARGARET A. EASTWOOD, Motorola, Inc., Schaumburg, Illinois
EDITH M. FLANIGEN, UOP (retired), White Plains, New York
JOHN W. GILLESPIE, University of Delaware, Newark
JAMIE C. HSU, General Motors, Warren, Michigan
RICHARD L. KEGG, Milacron, Inc., Cincinnati, Ohio
JAMES MATTICE, Universal Technology Corporation, Dayton, Ohio
CAROLYN W. MEYERS, North Carolina A&T State University, Greensboro
FRIEDRICH B. PRINZ, Stanford University, Palo Alto, California
DALIBOR F. VRSALOVIC, AT&T Laboratories, Menlo Park, California
JOSEPH WIRTH, RayChem Corp. (retired), Los Altos, California
JOEL S. YUDKEN, AFL-CIO, Washington, D.C.

RICHARD CHAIT, Director

Acknowledgments

The Committee on Defense Manufacturing in 2010 and Beyond would like to thank the following individuals for their presentations: John H. Bradham, South Carolina Research Authority; Lt. Col. Nina Brokaw, Defense Systems Management College; Todd Carrico, Advanced Research Program Agency; Andrew Dallas, Maritech; John A. DeCaire, National Center for Manufacturing Sciences; Sy Deitchman, U.S. Navy and Marine Corps; Gerald E. Ennis, The Boeing Company; Col. James Feigley, U.S. Marine Corps; Brig. Gen. Harry D. Gatanas, U.S. Army; Steven L. Goldman, Lehigh University; Beryl A. Harman, Defense Systems Management College; Robert Kiggans, Advanced Technology Institute; Lt. Col. Michael B. Leahy, Jr., U.S. Air Force; Steve Linder, Office of Naval Research; Lt. Gen. Les C. Lyles, U.S. Air Force; Don Meadows, Lockheed-Martin; Michael McGrath, Department of Defense; John Phillips, AlliedSignal Aerospace Services; Al Pruden, Jr., Lockheed-Martin; Herm M. Reininga, Rockwell International; Col. William F. Scott, Naval Aviation Depot, Marine Corps Air Station; L. Albert West, Sandia National Laboratories; Lt. Col. Earl Wyatt, U.S. Air Force.

This report has been reviewed by individuals chosen for their diverse perspectives and technical expertise, in accordance with procedures approved by the NRC's Report Review Committee. The purpose of this independent review is to provide candid and critical comments that will assist the authors and the NRC in making the published report as sound as possible and to ensure that the report meets the institutional standards for objectivity, evidence, and responsiveness to the study charge. The content of the review comments and draft manuscript remain confidential to protect the integrity of the deliberative process. We wish to thank the following individuals for their participation in the review of this report: Larry Cruzen, Cruzen Technologies, Inc.; Gary L. Denman, GRC International; James A. Jordan, Jr., consultant; Pradeep K. Khosla, Carnegie-Mellon University; Charles Lillie, Science Applications International Corp.; Herm M. Reininga, Rockwell International; Peter Sferro, Ford Motor Company; James Solberg, Purdue University; and Gen. William G.T. Tuttle, Jr., Logistics Management Institute.

Finally, the committee gratefully acknowledges the support of the staff of the Board on Manufacturing and Engineering Design, including Bob Rusnak, study director until October 1998, and Bonnie Scarborough and Tom Munns, who took over as study directors after October 1998. In addition, the work of the committee was greatly aided by Aida Neel, Charlie Hach, and Lois Lobo.

Contents

EXECUTIVE SUMMARY   1

1   NEW CONTEXT FOR DEFENSE MANUFACTURING   7
    Introduction, 7
    U.S. Defense Industrial Base, 11
    New Challenges for Defense Manufacturing, 12
    Committee on Defense Manufacturing in 2010 and Beyond, 16
    Manufacturing Technology Program, 17

2   DEFENSE MANUFACTURING CAPABILITIES REQUIRED FOR 2010   19
    Introduction, 19
    Defense Needs for 2010, 20
    Summary, 37

3   LEVERAGING ADVANCES IN COMMERCIAL MANUFACTURING   47
    Introduction, 47
    Advances in Commercial Manufacturing, 48
    Leveraging Commercial Advances, 62
    Summary, 69

4	NEW PRIORITIES FOR DEFENSE MANUFACTURING Setting Priorities, 73 Reorienting Programs, 77 Summary, 80	73

REFERENCES 83

APPENDICES

A	HISTORICAL PERSPECTIVE ON THE U.S. DEFENSE INDUSTRIAL BASE	87
B	WORLDWIDE WEB SITES AND DOCUMENTS RELATED TO DEFENSE MANUFACTURING	95
C	BIOGRAPHICAL SKETCHES OF COMMITTEE MEMBERS	99

Tables and Figures

TABLES

1-1 Pressures and Opportunities for Defense Manufacturing, 14

2-1 Required Defense Manufacturing Capabilities Based on the Defense Technology Area Plan, 39
2-2 Broad Categories of Required Defense Manufacturing Capabilities, 44

3-1 Commercial Manufacturing Advances and their Elements, 49
3-2 Defense Manufacturing Challenges Supported by Commercial Advances, 71

FIGURES

1-1 Defense budgets from 1962 to 2002, 9
1-2 Consolidation of the U.S. defense manufacturing industry from 1985 to 1995, 9

Acronyms and Abbreviations

ADA	software programming language (U.S. Department of Defense)
CAD	computer-aided design
CAM	computer-aided manufacturing
CAIV	cost as an independent variable
COTS	commercial, off-the-shelf (products)
DOD	U.S. Department of Defense
DTAP	Defense Technology Area Plan
FLIR	forward-looking infrared (sensors)
g	gravity
GNC	generative numerical control
GOCO	government-owned, contract-operated
HM&E	hull, mechanical, and electrical
HSM	high-speed machining
IPPD	integrated product and process development
IRST	infrared search and track (sensors)
LCD	liquid crystal display

ManTech	Manufacturing Technology Program
MLRS	multiple launch rocket system
MMIC	monolithic microwave integrated circuit
NDI	nondestructive inspection
RDT&E	research, development, test, and evaluation
SALT II	Strategic Arms Limitation Talks II
TOW	tube-launched, optically-tracked, wire-guided (missile)
VSA	variation simulation analysis

Executive Summary

Manufacturing[1] has played a vital role in the development and production of weapons systems used for the defense of the nation. During the Cold War, when defense manufacturing practices and capabilities evolved to meet specific threats to national security, defense products were manufactured largely by a dedicated defense industry. Since the end of the Cold War, however, changing circumstances have significantly influenced defense manufacturing. These include: changing threats to national security; declining defense budgets; consolidation of the defense industry; the increasing globalization of industry; the increasing rate of change of technology; and requirements for environmentally compatible manufacturing.

The National Research Council's Committee on Defense Manufacturing in 2010 and Beyond was formed to identify a framework for defense manufacturing in 2010 and to recommend strategies for attaining the capabilities that will be needed. To accomplish these objectives the committee (1) reviewed major trends that are changing the context of defense manufacturing and identified challenges to be met; (2) reviewed existing defense planning documents to identify defense-critical and defense-unique manufacturing capabilities; (3) reviewed advances in commercial manufacturing and identified those with the potential to meet defense manufacturing challenges, and (4) recommended strategies for developing the manufacturing capabilities that will be required in 2010 and beyond.

---

[1] For the purposes of this study, "manufacturing" has been broadly defined to include activities throughout the product life cycle (from needs assessment to concept formulation to production to disposal), as well as required resources (materials, infrastructure, information, people, time, money).

REQUIRED DEFENSE MANUFACTURING CAPABILITIES

After reviewing the technologies and manufacturing requirements described in defense planning documents and presentations by industry experts, the committee identified defense manufacturing capabilities required for 2010 that were defense-unique and/or defense-critical. These manufacturing capabilities, either broadly applicable to a number of weapons systems or specific to certain weapons systems, fall into the following six categories: composites processing and repair; electronics processes; information technology systems; weapons system sustainment[2]; design, modeling, and simulation; and production processes.

ADVANCES IN COMMERCIAL MANUFACTURING

The committee identified the following advances in commercial manufacturing as having the greatest potential for benefiting defense manufacturing: industry collaboration, adaptive enterprises, high-performance organizations, life-cycle perspectives, advanced manufacturing processing technology, environmentally compatible manufacturing, and shared information environments. These advances interact with each other and are composed of the following elements:

- **advanced approaches to manufacturing accounting**, including activity-based accounting and cost-as-an-independent-variable accounting
- **advanced approaches to product design**, including life-cycle design, integrated product and process development, three-dimensional digital product models, simulation and modeling, and rapid prototyping
- **advanced approaches to manufacturing processes**, including generative numerical control, adaptive machine control, predictive process control, high-speed machining, flexible tooling, soft tooling, tool-less assembly, embedded sensors, flip chips, nanotechnology, and biotechnology
- **environmentally compatible manufacturing technologies**, including cleaning systems, coatings, and materials selection, storage, and disposal
- **advanced approaches to business organization**, including teaming among organizations, virtual enterprises, long-term supplier relationships, high-performance organizations, cross-functional teams, lean enterprises, adaptive enterprises, agile enterprises, and knowledge-based and learning enterprises
- **information and communications technologies**, including electronic commerce, virtual co-location of people, data interchange standards, Internet technologies, intranet technologies, browser technologies, intelligent agents, seamless data environments, telecommunications, and distance learning

---

[2] For the purposes of this study, "sustainment" refers to the provision of personnel, logistics, and other support required to maintain operations until successful accomplishment of a mission.

NEW PRIORITIES FOR DEFENSE MANUFACTURING

Barring unforeseen international crises, defense budgets are unlikely to increase significantly in the near future. The committee believes that the principal criterion for prioritizing manufacturing capabilities should be potential cost savings (i.e., return on investment). Capabilities that meet this criterion are those that (1) will be applicable to many weapons systems or many elements of life-cycle costs, (2) will benefit from substantial nondefense resources, (3) will address large expenditure budget items for the Department of Defense, (4) could lead to significant performance or productivity gains, (5) will address problems likely to become more important in the future, or (6) will not be developed as a result of commercial investment.

Recommendation. Current Department of Defense research and development efforts in defense manufacturing should be augmented in four high-priority areas:

- efficient sustainment of weapons systems
- modeling and simulation-based design tools
- leveraging of commercial resources
- cross-cutting defense-unique production processes

Recommendation. Current and future Department of Defense research and development efforts aimed at improving manufacturing capabilities for sustainment of weapons systems should emphasize the following areas:

- application of advanced production processes and practices to maintenance, repair, and upgrade operations
- technology insertion for new and existing systems
- self-diagnostics for mechanical and electronic systems
- new technologies for remanufacturing
- design methods that improve sustainment

Recommendation. The Department of Defense should further encourage defense industry efforts to make the most of the simulation-based design environment and focus on the following activities:

- promote the development of models of defense products, manufacturing processes, and life-cycle performance
- develop algorithms for design trade-offs to optimize life-cycle costs
- develop enhanced and easily usable parametric models that facilitate design trade-offs at the conceptual stage
- initiate the development of product databases that will permit simulation at various levels of resolution

Recommendation. Advances in commercial manufacturing should continue to be monitored and adapted to defense applications as appropriate. Technology road maps created by commercial industry should be used to help defense manufacturing programs keep abreast of developments and forecasts.

Recommendation. The following development areas should be pursued to facilitate the widespread application of commercial, off-the-shelf (COTS) products:

- new weapons systems designed for open architecture and technology transparency
- a central program and mechanisms to maintain awareness of, document, and plan for new COTS technologies that can be incorporated into current and future weapons systems, as well as to disseminate this information to individual program offices
- improved methods of inserting COTS products in fielded weapons systems
- low-cost validation methods for determining the adequacy of COTS parts for military applications

Recommendation. Defense manufacturing programs should continue to address the development and improvement of defense-unique and defense-critical processes. The following defense-unique and/or defense-critical processes have the broadest range of applications:

- processes that enable rate-transparent production (i.e., production where the per unit cost is independent of the production rate)
- processes for the low-cost fabrication of composite structures
- processes for the low-cost production and application of coatings and structures with low observability
- defense-unique electronic technologies
- design, information, and manufacturing technologies that provide dimensional control in the production of large, complex parts

REORIENTING DEFENSE MANUFACTURING PROGRAMS

The Department of Defense Manufacturing Technology (ManTech) Program is a joint program of the armed services and the Defense Logistics Agency. The purpose of the ManTech program is to develop manufacturing technologies for the affordable, low-risk development and production of weapons systems. The current ManTech program has six thrust areas: metals processing and manufacturing; composites processing and manufacturing; electronics processing and manufacturing; advanced industrial practices; manufacturing and engineering systems; and sustainment/readiness. The committee believes that the ManTech program is an ideal vehicle for developing many of the required defense manufacturing capabilities described in this report and recommends the following ways in which the program can be reorganized to meet future demands.

Recommendation. The ManTech program should play the following roles in the 2010 time frame (some of these roles require only a change in emphasis of existing roles; some are new roles that should be incorporated into the program charter):

- *Leader in affordability.* The ManTech program should be considered the primary means of achieving affordability throughout the life cycle of weapons systems.
- *Focal point for cross-cutting defense technologies.* The ManTech program should focus on projects whose results are expected to be widely applicable.
- Technology middleman. The ManTech program should aggressively promote the implementation and dissemination of new technologies.
- *Information broker and planner.* The ManTech program should expand its role in providing information on new technologies to the defense community.
- *Expert in weapons systems technologies.* The ManTech program should provide expertise in the technologies important to major weapons systems.

Recommendation. The ManTech program should consider revising its division of effort if it is to implement the new roles and development initiatives that the committee has recommended. The following changes are recommended:

- *Production processes.* This area should remain a major thrust area, but the emphasis should be shifted toward cross-cutting technologies.
- *Advanced industrial practices.* This area should be expanded beyond industrial best practices to include technologies for enhancing cost-effectiveness.
- *Manufacturing and engineering systems.* The ManTech program should establish an initiative for the development of simulation-based design tools.
- *Sustainment of weapons systems.* This area should be greatly expanded and should be given as high a priority as production processes.
- *Leveraging of commercial resources.* The ManTech program should establish an initiative for leveraging commercial resources with an emphasis on COTS products.

The committee believes that the ManTech program could be reoriented without compromising the important initiatives already under way. Investments in the ManTech program already provide a return through cost savings and cost avoidance. The recommended emphasis on projects and technologies with broad applicability should ultimately increase the return on investments.

1

New Context for Defense Manufacturing

INTRODUCTION

Defense manufacturing[1] —a keystone of the nation's security—is undergoing sweeping changes. Changing threats to national security, declining defense budgets, the consolidation of the defense industry, the globalization of industry, the increasing rate of change of technology, and requirements for environmentally compatible manufacturing are all contributing to new challenges being faced by defense manufacturing.

Superior combat capability is achieved in part by the use of cutting-edge technologies in weapons systems. Manufacturing systems are required to successfully transform designs for high technology weapons systems into operational products. Since its establishment in 1947, the U.S. Department of Defense (DOD) has played various roles vis-à-vis the manufacturing sector, including customer, end user, service provider, co-developer and co-producer of products, and co-funder and co-manager of both basic research and technology research and development. The new context for defense manufacturing will result in a transformation of the relationships between defense and commercial sectors and will require that new priorities be set for the allocation of defense manufacturing resources.

---

[1] Defense manufacturing is defined as manufacturing activities that produce defense-related products. For the purposes of this study, "manufacturing" has been broadly defined to include activities throughout the product life cycle (from needs assessment to concept formulation to production to disposal), as well as required resources (materials, infrastructure, information, workers, time, money).

Changing Nature of Threats to National Security

The nature of threats to U.S. national security are substantially different today than they were during the Cold War. For example, during the Cold War, defense policy focused on global conflict, and the engagement scenarios considered most likely involved nuclear weapons. Current defense policy is focused on regional conflicts with engagement scenarios involving conventional weapons. National policy requires that U.S. military forces maintain a credible nuclear deterrence, as well as the ability to engage in and prevail in one and one-half "contingency" or regional operations involving combat against enemy forces but no global political or global military implications.

In addition, U.S. military forces will continue to be committed to a variety of operations, known as "military operations other than war." These missions, which are different in character and scope from traditional military activities, include providing support for drug interdiction and international peacekeeping. These activities present unique challenges and risks, as well as new demands on weapons and logistic systems. For example, military forces and equipment normally available for contingency operations, training, or maintenance might not be available during nontraditional military deployments.

Declining Defense Budgets

For the past decade, the resources available to DOD have been declining (see Figure 1-1). In 1997, the DOD budget was $258 billion, down from $382 billion in 1989. Projections to the year 2003 show a modest increase from current levels (OMB, 1998). Barring a major international crisis, however, defense budgets are not likely to grow significantly in the next decade.

The decline is even more pronounced in terms of the development and production of new weapons systems. The procurement budget has been reduced from $106 billion in 1989 to $48 billion in 1997, while the operations and maintenance budget has remained stable at about $100 billion. Reductions in spending on engineering and manufacturing development also indicate that procurement programs for new weapons systems will be scarce for the next 10 years. Although the research, development, test, and evaluation (RDT&E) budget has remained relatively flat in constant dollars since 1989, this represents a loss of buying power of about 30 percent. According to DOD projections, the RDT&E budget will decrease by about 6 percent over the next five years in terms of actual dollars (OMB, 1998).

Consolidation of the Defense Industry

The defense industrial base has historically consisted of companies responsible for most of the research, development, and manufacture of defense systems.

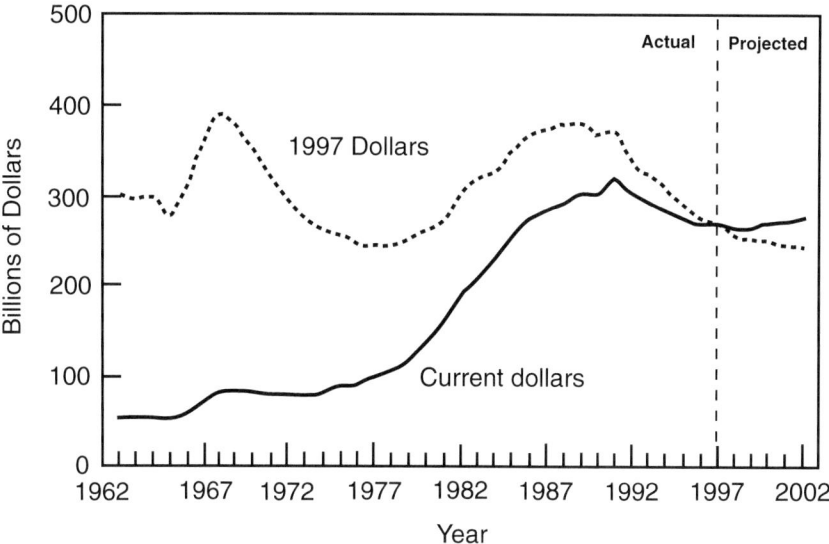

FIGURE 1-1 Defense budgets from 1962 to 2002 (in 1997 dollars). Source: Congressional Budget Office, 1997.

Since the late 1980s, these companies have been involved in a wave of mergers and consolidations (see Figure 1-2). As a result, only three major defense contractors remain in the United States today (Boeing, Lockheed Martin, and Raytheon), and these three are increasingly turning to commercial markets. In fact, commercial sales currently outweigh defense sales for all of them (Defense Science Board, 1997).

Globalization of Industry

Manufacturing is becoming an increasingly integrated global system as a result of several factors, including: the growth of emerging economies, the formation of companies of indistinct nationality, the dispersal of design and production capacity, and the outsourcing of design and production. The global economy is becoming highly integrated with information, funds, materials, components, final products, and workers crossing national and regional boundaries daily. Companies can design and manufacture products in many different locations in the world; manufacturing capacity has become a commodity (NRC, 1995).

According to Brooks and Guile (1987), these changes have implications for the role of the United States in the global economy. Although the United States is no longer the dominant player, the country now has access to markets that were unreachable only a decade ago. Business leaders have become accustomed to

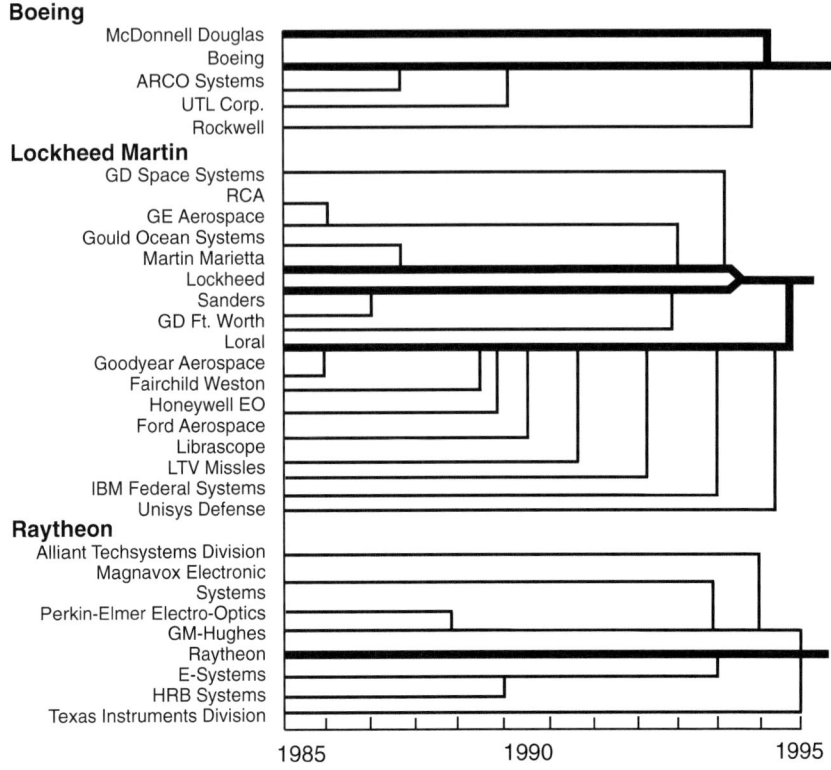

FIGURE 1-2 Consolidation of the U.S. defense industry from 1985 to 1995. Source: Dowdy, 1997.

thinking of the global marketplace as a significant source of consumer demand, as well as an attractive source of supplies, including foreign technology, intellectual capacity, manufacturing capacity, and investment capital, as well as traditional imports, such as natural resources and hard goods. The logistics of currency exchange, information exchange, and transportation are well understood, and facilitators of global trade are available. The trend toward collaboration on a global scale is part of the U.S. management culture, and companies often join with foreign partners to augment their capabilities.

Increasing Rate of Change of Technology

Several manufacturing technologies, including biotechnology, nanotechnology, electronic and microelectronic technologies, and information technology, are on the verge of revolutionary development. Changes in microelectronics have been particularly rapid with the minimum feature size of a 256-kbit dynamic random access memory microchip decreasing from 1.5 µm to 0.25 µm between 1983 and 1997. Soon, 1-Gbit dynamic random access memory microchips with feature sizes of 0.18µm will be available (Rothschild et al., 1997). Advances in information technology have affected and will continue to affect all aspects of manufacturing, including product and process design, shop floor controls, modeling and simulation, enterprise integration and business practices, communications, social structures, and education and training (Dertouzos et al., 1989; Friedman et al., 1992).

Requirements for Environmentally Compatible Manufacturing

Expectations of responsible behavior by industry and government have been rising with regard to environmental concerns. This trend is evident in national and international efforts to mitigate environmentally harmful effects of industrial processes and to improve the decision making for handling and disposing of industrial contaminants. Manufacturing as a whole is being driven to convert to environmentally benign processes (NSTC, 1997) and will continue to be affected by environmental regulations. Commercial industry will have to develop the necessary process technologies and practices, and defense manufacturing will have to reduce environmental effects of depot and field operations, as well as weapons disposal. Life-cycle design processes will have to include disposal costs and incorporate design features to mitigate them.

U.S. DEFENSE INDUSTRIAL BASE

Manufacturing the products required by the U.S. armed services has traditionally been the responsibility of the "defense industrial base," industrial and military facilities devoted partially or entirely to the production of defense-related products. The size and character of the defense industrial base has changed many times since the nation was founded. Just after the War of Independence, the U.S. defense industrial base consisted of purveyors of powder and guns, storage locations for weapons and supplies, and state-owned shipyards. In 1950, the defense industrial base was formally defined by the Defense Production Act, which also established a priority system for obtaining military hardware and software during emergencies and provided seed money for improving manufacturing facilities and processes. The defense industrial base remained clearly defined throughout most of the Cold War and the conflict in Southeast Asia (see Appendix A for more details).

The 1990s have been a decade of profound change for both the prime contractors of the defense industrial base and their suppliers. The boundaries between the defense industrial base and commercial industry have become increasingly difficult to determine. Although numerous aircraft plants, arsenals, and shipyards are still devoted primarily or exclusively to producing military hardware, many now assemble systems from components manufactured elsewhere, usually by commercial industry. These changes are as dramatic as the changes during the build-up for World War II.

NEW CHALLENGES FOR DEFENSE MANUFACTURING

The major forces for change impacting defense manufacturing interact with each other and with the defense and manufacturing sectors to create both new pressures and new opportunities. Meeting the new threats to national security will require reconfigurable weapons systems, production surge capacity, and access to production sources. Overall, declining defense budgets and increasing systems complexity will result in fewer new weapons systems being developed and produced, as well as fewer weapons systems of any kind being procured. This trend toward fewer procurements will create a need for low-volume production processes and systems at a reasonable cost. At low production volumes, nonrecurring costs, driven largely by development costs and cycle times, are greatly magnified on a per unit basis, and reductions in both recurring and nonrecurring costs will be necessary. Budget constraints will also increase the demand for low-cost weapons systems that have favorable life-cycle costs and benefits.

The lives of existing weapons systems will be extended. Currently, many aircraft in the operational force are more than 20 years old, and weapons systems for which no new replacements are planned are expected to remain in service an additional 25 years (NRC, 1997). The sustainment[2] of aging weapons systems will require effective, low-cost maintenance in the field and at depots and maintenance facilities, which will require significant reductions in cycle time to reduce the costs. Cost-effective methods will also be necessary to upgrade aging systems and remanufacture spare parts.

Due to reduced expenditures, the ability of DOD to influence industry will decline. During the 1980s, the defense industry, in response to demands from DOD for lower costs, vectored significant independent research and development funding into manufacturing related areas. However, in the 1990s, this investment has been decreased by as much as 70 percent because of reductions in military procurement budgets, and there has been a shift from funding of independent research and development to funding of bids and proposals, as the industry has increased its focus on winning bids. The loss of competition as a result of industry consolidations may significantly reduce the incentives for industry to invest in cost reduction measures. Defense consolidation has vastly diminished the flexibility required for surge capacity (due to plant closings), diminished competitive

---

[2] For the purposes of this study, "sustainment" refers to the provision of personnel, logistics, and other support required to maintain operations until successful accomplishment of a mission.

innovations in products, and reduced competitive pricing based on multiple sources for products.

The decreasing RDT&E budget has reduced DOD's ability to develop new technologies for effective and affordable defense manufacturing. Defense manufacturing will, therefore, have to carefully prioritize spending in priority areas and try to leverage innovations in commercial manufacturing by maintaining its awareness of developments, transferring new technologies to defense applications, and investing in the adaptation of commercial technologies to defense applications.

Industry reacts predictably to fluctuations in defense spending. When budgets are rising or at peak levels, industry invests in defense-related infrastructure, equipment, and processes. When budgets are declining or at low levels, investments of all types are decreased or directed toward other markets. Defense authorities have always been able to mobilize the U.S. commercial manufacturing industry in times of national crisis. In peacetime, however, the influence of national defense requirements on the priorities of commercial industry are limited. Defense agencies were once able to dictate industrial directions even though commercial interest in these technologies was low. In the future, however, defense agencies will only be able to justify programs that meet defense-unique requirements, and the investment will have to come from the government.

On the one hand, the globalization of industry has resulted in greater access to new processes and technologies that can improve the performance and quality of defense products while reducing costs. On the other hand, greater reliance on foreign sources could threaten the security of product information and, in times of conflict, product sources. If industry alone decides where products will be designed and manufactured, protections for U.S. defense technologies and capabilities will have to be established. If weapons used by U.S. forces are produced and supported by foreign industrial organizations or governments, then DOD will have to determine which products and processes must be protected and develop either alternative sources for critical foreign suppliers and/or rapid reverse engineering and remanufacture capabilities. DOD and its prime contractors will have to monitor the actual sources used by suppliers, either through guidelines to prime contractors that dictate where critical defense parts can be manufactured or through requirements that prime manufacturers be responsible for the life-cycle support of their products.

New technologies create new opportunities. Technological advances can be used to upgrade weapons systems or to develop completely new capabilities. Weapons systems upgrades will most likely be accelerated, and new capabilities in weapons systems will be introduced. Future advances in information technology and applications of information technology to manufacturing can also be applied to defense requirements.

Table 1-1 summarizes the forces that are changing the context of defense

Table 1-1 Pressures and Opportunities for Defense Manufacturing

Force for Change	Effect on Defense	Resulting Pressures and Opportunities
Changing nature of threats to national security	Increased variety of military missions Increased unpredictability of military missions Increased emphasis on conventional rather than nuclear weapons	Capability for customizing weapons systems Manufacturing surge capacity Design and manufacture for multiservice use Design for reconfigurability Rapid product realization
Declining defense budgets	Demand for affordable weapons systems Fewer new weapons systems procured and produced in lower volumes Extension of weapons systems life	Reduction in development cycle times and costs Application of cost-as-an-independent-variable accounting Design and manufacture for multiservice use Use of commercial-off-the-shelf products Design for low life-cycle costs Low-cost processes for low-volume production Low nonrecurring costs in product realization Reduction in cycle times Extended life in new systems Design for maintainability Design for technology insertion Efficient maintenance and depot operations Remanufacturing processes Improved diagnostics Product and process databases
Consolidation of the defense industry	Potential for reduced capacity, competition, and innovation Potential for lower priority given to defense production	Processes and systems for surge production Strategies for maintaining innovation and competition Production of defense systems on commercial production lines

NEW CONTEXT FOR DEFENSE MANUFACTURING

Table 1-1 continued

Force for Change	Effect on Defense	Resulting Pressures and Opportunities
Globalization of industry	Location of component and subsystem development, design, and production determined by industry Greater reliance on foreign suppliers Security threat for product information Potential loss of production sources in time of conflict Increased access to new product and process technology	Guidelines on critical components and subsystems Identification of suppliers Development of security systems for product and process data Remanufacturing capability for components and subsystems Alternate sourcing strategies Adapt commercial "best practices"
Increasing rate of change of technology	Opportunity for more frequent improvement of existing weapons systems Opportunities for introduction of new weapons capabilities and new weapons systems Potential for improved product databases, program management, and retention of production know-how	Open-architecture systems to enable technology insertion Reduced product realization time Use of industry road maps in product development and design Adaptation of technologies to defense-specific applications Development of product and process databases
Requirements for environmentally compatible manufacturing	Stringent environmental regulations for manufacturing and maintenance and depot operations	Life-cycle design Environmentally compatible production processes Reduced pollution in depot and maintenance operations

manufacturing and the resulting pressures and opportunities. These pressures and opportunities can be grouped into the following categories:

- low-cost rapid product realization
- expanded design capabilities
- environmentally compatible manufacturing
- adaptation of information technology
- security of product and process data
- access to production sources
- use of commercial manufacturing capacity
- sustainment of weapons systems

COMMITTEE ON DEFENSE MANUFACTURING IN 2010 AND BEYOND

The National Research Council's Committee on Defense Manufacturing in 2010 and Beyond was formed in response to a request by the armed services[3] that a framework for defense manufacturing in 2010 be identified and that strategies for attaining the necessary capabilities be recommended. The committee was asked to (1) review major trends that are changing the context of defense manufacturing and identify challenges to be met; (2) review existing defense planning documents to identify defense-critical and defense-unique manufacturing capabilities; (3) review advances in commercial manufacturing and identify those with the potential to meet defense manufacturing challenges; and (4) recommend strategies for developing the manufacturing capabilities required for the year 2010 and beyond. This report reflects the results of the committee's activities.

Methodology and Report Organization

The committee notes that the issue of forecasting the future, even from existing trends, is complex. Trends that are clearly visible today may rapidly change, and projections of technology advances may turn out to be either optimistic or pessimistic in light of actual events. Each trend is an aggregate of projections derived from the perceptions of many individuals. The availability of a predicted technology advance will depend on the actual timetable of discoveries and engineering applications and the probability that alternative technological mechanisms will emerge.

To evaluate the needs of defense manufacturing in the year 2010 and beyond, the committee, therefore, relied on a variety of information sources. First, the committee reviewed requirements for defense manufacturing for the year 2010 and beyond, including DOD's Defense Technology Area Plan (DTAP), and asked representatives of defense organizations, prime contractors, and program officers to present their future needs. Based on this material, the committee identified "defense-critical" manufacturing requirements, as well as "defense-unique" requirements (i.e., areas unlikely to attract significant commercial investment). The committee then identified required manufacturing capabilities (described in Chapter 2).

Second, the committee reviewed forward-looking manufacturing studies (e.g., Next Generation Manufacturing [NGM, 1997] and Visionary Manufacturing Challenges for 2020 [NRC, 1998]), reviewed information sources available on the World Wide Web (see Appendix B), and invited speakers to assess

---

[3] The study is sponsored by the U.S. Air Force, the U.S. Army, the U.S. Navy, and the Defense Logistics Agency.

advances anticipated in manufacturing. The committee summarized the current state of commercial manufacturing and identified several general advances in manufacturing that defense manufacturing can expect to draw on. These are described in Chapter 3, which also contains the committee's analysis of the extent to which these advances will meet defense manufacturing needs.

Finally, the committee developed criteria for setting priorities and identified the categories of manufacturing capability that could best meet these criteria. In addition, the committee developed recommendations on how DOD could develop the necessary manufacturing capabilities and identified new roles and priorities for the ManTech (Manufacturing Technology) Program. These conclusions and recommendations are presented in Chapter 4.

MANUFACTURING TECHNOLOGY PROGRAM

Throughout the nation's history, the armed services have strongly supported new technologies and manufacturing methodologies. In the late 1950s, DOD established the ManTech Program under the provisions of the Defense Production Act of 1950 and its extensions. The objective of the ManTech program was to strengthen the U.S. defense industrial base by encouraging the development and use of innovative manufacturing methods and processes. The program was based on the premise that in manufacturing technology areas where the cost to develop and implement an innovative production methodology would not be a prudent business risk, DOD should invest in bringing these ideas to fruition. It was assumed that, if DOD funding was available to augment private investment, contractors would be encouraged to upgrade their manufacturing facilities and processes and that the overall result would be high quality weapons systems that could be produced and delivered in shorter times and at lower cost.

The committee, although not explicitly asked to provide recommendations for the roles of the ManTech program, believes that ManTech is the logical organization through which many defense-critical and defense-unique manufacturing requirements should be developed. No other organization within the DOD has ManTech's history of research and development or a support structure already in place from the three branches of the armed services.

Each branch of the armed services has its own ManTech program and, during the first two decades of the program, there was little or no coordination between them. The Army applied its funds to improving processes used in the manufacture of various Army commodities and weapons systems. The Navy invested in the establishment of centers of excellence. The Air Force invested in large-scale, enterprise-wide information systems and technologies and improvements in materials processes. In 1975, the Secretary of Defense directed the armed services to increase their emphasis on and support of the ManTech program. A tri-service Manufacturing Technology Advisory Group was established to coordinate plans with industry associations. The review process created by the advisory group reduced, but did not eliminate, the duplication and overlap of projects. In the late 1970s, the joint logistics commanders (the Commander, Air Force System Command; Chief of Naval Materiel; and Commander, Army Materiel Command), acting as a joint body, forced the complete coordination of the ManTech program.

At that time, some ManTech funds were allocated to the adaptation of commercial products for use by the military in the nondevelopmental item program, the forerunner of the dual-use program. In the early 1980s, the Defense Science Board recommended that the ManTech program be funded at 2 percent of the procurement budget, but this level of funding was never reached. Beginning with fiscal year 1991, the ManTech program was gradually incorporated into DOD's Manufacturing Science and Technology program, where it remains today. Funding for the ManTech program originally came from the budget for RDT&E, although many people in DOD and Congress believed that it should be funded by procurement dollars.

In spite of varying levels of consolidation and support, the ManTech program has been responsible for a number of successes, including the following:

- the development of the first numerically controlled machine tool
- the establishment of automatically-programmed tools as an industry-wide standard language
- the establishment of processes and tools that have accelerated the development of the microelectronics industry
- the development of isothermal forging for net-shape manufacturing of titanium and superalloy parts
- the development of the first three-dimensional nondestructive inspection system for rocket motors and other critical parts
- the development of process modeling methods, such as integrated definition
- support for accelerating the development of computer-aided design, computer-aided manufacturing, and computer-integrated manufacturing
- significant improvements in night vision systems and capabilities

Although the F-16 technology modernization program was not part of the ManTech program and did not have the ManTech objective of improving manufacturing equipment and processes across industry, this program represents another notable success for ManTech. The objective of the program was to reduce F-16 unit procurement costs, and it was initially funded completely by procurement funds. As the program progressed, the contractor, General Dynamics, became convinced that benefits would accrue to the ManTech program from a similar approach. General Dynamics entered into an arrangement with the federal government whereby both of them provided funding to implement manufacturing improvements, and they shared the cost reductions according to the ratio of funds contributed. The program's achievements surpassed the expectations of its most ardent supporters, and the secret to its success was the sharing arrangement.

2

Defense Manufacturing Capabilities Required for 2010

INTRODUCTION

Historical events have always influenced the United States' perception of its defense needs, which, in turn, have influenced the nation's commitment to maintaining defense manufacturing capacity and supplies. Threats to national security today are substantially different, although no less demanding, than those of the Cold War period. Currently, defense policy focuses on regional rather than global conflicts that will most likely involve conventional rather than nuclear weapons. Therefore, conventional weapons and related manufacturing systems are being given higher priority for research, development, and procurement funding than nuclear systems.

Although many nuclear systems have been removed from alert status, decommissioned, or destroyed as required by the Strategic Arms Limitation Talks II (SALT II) and other agreements with the former Soviet Union, the United States maintains a nuclear deterrent. As long as Russia, China, or other potential adversaries have a nuclear strike capability, U.S. policy will require that a credible nuclear retaliation force be maintained. Strategic nuclear forces, although significantly reduced in size, must remain reliable and effective. Modifications and upgrades to the residual nuclear-capable force must continue as needed to maintain their readiness and effectiveness.

Today's force structure, personnel, and military equipment have been adapted to fulfill stated national policy requirements. Weapons systems designed and produced today must meet a broader range of mission requirements and must be reconfigurable in the field. This requires improved design methods and manufacturing processes, and manufacturing systems that enable the customization of weapons systems. In addition, manufacturing facilities must have production surge capacity to make rapid modifications and provide additional inventories in times of crisis.

DEFENSE NEEDS FOR 2010

The committee reviewed DOD documents to determine the technology areas that will be important for future defense manufacturing, including a DOD technology forecast regularly generated in an effort to plan for future needs. This forecast, known as the Defense Technology Area Plan (DTAP), presents DOD's objectives and investment strategies for the technologies considered to be critical to its acquisition plans, war-fighting capabilities, and joint war-fighting needs. The DTAP contains defense technology objectives in 10 technology areas: air platforms, ground and sea vehicles, weapons, sensors, electronics and battlefield environment, information technology systems, materials processes, chemical/biological defense and nuclear systems, biomedical systems, and space platforms (DTAP, 1997).

Defense-Critical and Defense-Unique Technologies

Product and process technologies are classified by DOD as "defense-critical," "defense-unique," or both, and they span the industrial technology spectrum. Whether or not a product technology is considered defense-critical depends on its applications. A manufacturing process technology is considered defense-critical if it is required to produce a defense-critical product. For example, microprocessors and other electronic technologies, which are indispensable to many defense systems, are defense-critical. A product or process technology is considered defense-unique if it is used only for defense purposes and has no commercial application. Therefore, a defense-critical technology may or may not be defense-unique, whereas a defense-unique technology is always defense-critical.

There are literally hundreds of technologies, subsystems, and systems that could be characterized as either defense-critical or defense-unique. The committee made no attempt to list them all or to describe them all in detail. Instead, the committee analyzed the range of DTAP defense-critical technologies under consideration and concentrated on those that were either defense-unique and, therefore, not likely to be developed by commercial industry, or those that had defense-unique applications, although the technology itself was not defense-unique. The committee then selected the technology areas that were most dependent on advances in manufacturing.

Weapons Systems Platform Technologies

Aircraft Weapons Systems

Weapons systems for air warfare, such as long-range bombers and maneuverable supersonic stealth fighters, will be necessary for the foreseeable future. Systems capable of carrying and employing a variety of weapons and of finding, engaging, and defeating enemy air, land, and sea forces and targets require many defense-unique technologies and manufacturing capabilities not available in the commercial world.

Although all air weapons systems require some cross-cutting technologies with defense-unique applications, only fighters and bombers require defense-unique technologies. Military transports, trainer helicopters, and utility aircraft require essentially the same technologies as their civilian counterparts. Because many military aircraft platforms are expected to be long-lived, several manufacturing considerations have become significant including: repair techniques for aging aircraft and nonintrusive, real-time methods for monitoring flight loads and damage (DTAP, 1997).

High g Loads and Acceleration. An aircraft's ability to sustain loads of eight to nine times the force of gravity (g) and to accelerate rapidly at high speeds requires high structural strength, unusual aerodynamic characteristics, a high thrust-to-weight ratio, and protecting the crew from the high g loads. A key manufacturing capability is the design and processing of high strength-to-weight materials, particularly composites, for which new design concepts and processing methods are needed to reduce costs. Electronic systems must also be designed and packaged to withstand the high g forces and vibrations in this severe environment.

Weapons Containment. The ability to carry guns, air-to-air and air-to-ground missiles, rockets, and bombs is a defense-unique requirement for aircraft. Bombers generally carry loads internally in a large fuselage designed to accommodate them; fighters generally carry loads externally on suspension and launch equipment. Different types of launch equipment include multiple-ejector and triple-ejector racks for bombs and ejector launchers for some air-to-air missiles (e.g., the Sparrow) on F-14, F-15, and F-18 fighters. Other air-to-air missiles (e.g., the Sidewinder) are rail launched. All types of launch equipment have two things in common: (1) they add a significant amount of drag to the aircraft, and (2) they have a large radar signature. Research and development in recent years has focused on stealth racks that reduce drag and have a smaller radar signature. With state-of-the-art manufacturing techniques, these racks could be designed and produced, but they would be expensive and would still compromise aerodynamics and stealth performance. Breakthroughs will be necessary to overcome these problems.

Stealth fighters, such as the F-117, F-22, and B-2, carry weapons internally, which reduces compromising the radar signature and does not as severely affect aerodynamic performance. But fighters with internal weapons bays require thicker fuselages and are, therefore, more expensive. The design and low-cost processing of high strength-to-weight materials would be an important step in solving this problem. The acoustic environment in a weapons bay with the bay door open is notoriously severe, particularly at high speeds, which has design and production implications for the weapons carried, particularly their electronic components. Another defense-unique requirement is that aircraft be able to eject munitions from the weapons bay at very high speeds and g loadings.

Surface and Subsurface Sea Combat Vessels

Defense-unique challenges for surface sea combat vessels include: reducing topside weight and volume while reducing signature and increasing sensor performance; minimizing the weight and volume of hull, mechanical, and electrical (HM&E) systems while increasing combat tolerance and decreasing life-cycle costs; improving damage fight-through and recovery while minimizing crew size and equipment redundancy; and developing automated intelligent monitoring and control systems for HM&E equipment. In terms of manufacturing, the overall requirement is affordability, which translates into new design concepts and low-cost processes, as well as effective sustainment techniques.

Defense-unique needs for subsurface sea combat vessels (submarines) include: reducing acoustic signatures and increasing shock resistance while reducing costs. These will require new system-level design approaches. Simulations during the design could accurately assess performance and couple cost data to high-level system designs, enabling necessary trade-offs.

Land Combat Vehicles

Land combat vehicles include the M-1 Abrams tank family, the Bradley fighting vehicle system, and the M-113 vehicle family. Some of these vehicles were first designed and produced several decades ago and, although they have been upgraded, most vehicles are at least 15 years old. Because they are expected to remain in the inventory for many years to come, their continued sustainment and tactical effectiveness must be ensured.

New vehicles planned for acquisition prior to 2010 include the future main battle tank; the future scout and cavalry system; the reconnaissance scout vehicle; the future combat system; the future infantry vehicle; and the advanced amphibious assault vehicle. All of these new vehicles will feature innovative technologies. Some of the technologies being considered for new vehicles and for upgrading existing vehicles are better weapons, advanced armor, new lightweight materials, composite structures, semi-active suspension, advanced propulsion systems, electric drive, new electronic architectures, intervehicle and intravehicle digitization, an intravehicle electronics suite, advanced crew station technology, new methods of signature suppression, fire suppression systems, laser protection, hit avoidance techniques, active protection systems, and new turret technologies.

The broad range of technology requirements for land warfare systems are related to meeting deployment requirements while increasing survivability and lethality. Requirements include: smaller crews, automated drivers, training, smaller radar signatures, reduced mobility component weight and volume, and increased power. Integrated product and process development and virtual prototyping are two of the manufacturing capabilities that will be critical to meeting these challenges (DTAP, 1997).

Weapons Technologies

Expendable Munitions

The military services use hundreds of different types of expendable munitions. The term "munitions" is used here to refer to expendable ordnance other than large missiles (e.g., fuzes, land and sea mines, and aimable warheads). The production of munitions, in this narrow sense of the term, requires technologies and manufacturing capabilities that are defense-unique (i.e., they generally have no commercial application) with the exception of ammunition for personal and law enforcement weapons and explosives used in mining and construction. Although all expendable munitions have specific requirements, some requirements are common to all of them, including quantity, quality, long-term storability, and affordability. Critical technologies in the munitions field are generally related to the safety and efficiency of manufacturing processes, affordability, storage and handling, the effectiveness of hydrocodes and warheads, sensors, arming and fuzing, and methods of tactical delivery or deployment. A high-yield, robust process for fuze production is required.

The manufacturing process of filling munitions with explosive materials, called "load, arm, and pack," is usually done at military arsenals, depots, or government-owned, contractor-operated (GOCO) plants located at arsenals (although filling of ammunition up to 30mm is often done in contractor-owned facilities). A key requirement is precision filling of the explosive (i.e., filling without voids). In some cases, costly 100 percent inspection is required. Precise metering methods could bring down this cost. New automated processes for filling munitions with explosive materials could minimize human intervention, promote safety, improve process yield, and ensure performance.

Missiles and Torpedoes

Missiles and torpedoes have no commercial application, with the exception of space launch vehicles, and can therefore be considered defense-unique technologies. Compared to expendable munitions, missiles and torpedoes are more expensive to manufacture, more capable, and are produced in smaller quantities.

Manufacturing requirements for missiles identified in DTAP include: efficient packaging of all components in a missile the size of a tube-launched, optically-tracked, wire-guided (TOW) missile; the development, design, and integration of miniaturized guidance and control actuators with an advanced composite propulsion system in a small-diameter hypervelocity missile; a low-cost, small, producible, strap-down mechanism and guidance components for precision guidance of a highly rolling small rocket; the design of shipboard launch systems that can accommodate a wide range of missiles; the incorporation of attachments in missile airframes constructed of composite materials that do not compromise operational capability; low-cost, lightweight-composite external surfaces that meet the high temperature and stiffness requirements of a tactical missile; improved strength-to-weight/volume ratios and reduced erosion and weight of insulation (for solid propellant rockets); and smaller ramjet components. These technical challenges in manufacturing requirements are aimed at miniaturization, low-cost production processes, and advanced composite materials and processes and should be addressed in an integrated manner.

Many of the requirements and challenges identified for missiles also apply to torpedoes. Undersea weapons, however, also have their own manufacturing challenges, including a 40 percent reduction in development and ownership costs by 2005 and the use of more than 50 percent common subsystems by 2010. The first of these challenges requires reductions in development cycle time, reductions in nonrecurring costs, and the development of efficient sustainment methods. The second challenge requires overall system designs based on common subsystems.

Guns

The small arms industry sells more guns commercially than it does to the military, but the remainder of the gun industry is basically defense-unique. Artillery tubes, mortars, machine guns, fully automatic and large-caliber personal weapons, armored vehicle guns, naval guns, and their associated aiming and loading mechanisms have no civilian counterparts, although law enforcement agencies use certain types of fully automatic personal weapons. Military guns are not produced in the same quantities as expendable munitions, but most are produced in quantities ranging from thousands to tens of thousands. Cycle times and nonrecurring costs will have to be reduced.

Mobile Weapons Systems

Mobile and transportable crew-operated guns, rockets, and missile systems include crew-operated machine guns; self-propelled and towed artillery, howitzers, and rockets; and missile systems. Specific examples are the Paladin, the Crusader, the advanced tactical missile system, the multiple launch rocket system (MLRS), and the TOW missile. New subsystems that can be inserted into existing systems to upgrade their capability and sustainability include the XM982 extended range 155mm artillery projectile, the XM297 Crusader solid propellant cannon, the XM291 120mm tank gun, and the electrothermal chemical version of the XM291 tank gun. New systems on the horizon include the high mobility artillery rocket system as a replacement for the MLRS, the guided MLRS, a countermissile rocket launcher, a follow-on to the TOW missile, the objective-crew-served weapon, and the extended range guided missile. Longer-range technologies in this area include electromagnetic guns and directed energy weapons. Several major manufacturing and design challenges are associated with these weapons, including: packaging constraints for electrothermal chemical technologies; the development of high-efficiency plasma ignitors; the development of high-energy-density propellants; the development of an advanced medium-caliber composite barrel with high-efficiency rail design; weight minimization; and smaller component sizes for electromagnetic and directed energy weapons.

Cross-cutting Technologies

Several technology areas for defense products, which are broadly applicable to defense systems, are discussed below. These cross-cutting technologies include: low observability techniques, sensors, electronics, and information systems.

Low Observability Techniques

The need for low observability, or "stealth," is clearly a defense-unique requirement that has no meaningful commercial counterpart. Although the public usually associates stealth with aircraft, such as the F-117, F-22, and B-2, the need for stealth is a characteristic of all weapons and weapons systems used by the armed forces. Stealth will be a prominent performance characteristic in the design of many new manned and unmanned vehicles, missiles and weapons, and other equipment. As potential adversaries develop the means to counter low observability technology, designers and manufacturers of defense systems must find ways to reduce observables even further. New designs for achieving low observability must be affordable and must address the issues of manufacturability and supportability, as well as stealth performance. Manufacturing tools, methods, and practices may have to be modified to accommodate the affordable production of equipment.

Systems with Stealth Requirements. Personal dress and accoutrements for soldiers and marines have been designed for low observability. Camouflage-patterned clothing and helmets and grease paint for faces and hands have been used for many years. Personal and crew-operated weapons must also be designed for low observability in order to enhance survival, as well as combat effectiveness. Reduced muzzle and rocket blasts, as well as reductions of other telltale signature elements, all contribute to stealth.

Ships need reduced radar, infrared, visual, and acoustic signatures. The latest Nimitz class aircraft carrier (CVN-77), scheduled to be delivered to the Navy in 2008, will have significantly lower observability than its predecessors, including a redesigned and smaller island, fewer angular protuberances, and rounded deck edges. The CVX carrier, scheduled to enter the fleet in 2013 to replace the U.S.S. *Enterprise* (CVN-65) built in 1965, will be a new design from the keel up. Low observability will be one of the principal design characteristics of the CVX, which is expected to be in service throughout most of the next century.

For many years, the 688 Los Angeles class submarine was the world's quietest attack submarine; however, according to the Office of Naval Intelligence, the newest Russian submarines, the improved Akula boats, are quieter (ONI, 1995). The commissioning of SSN-21 *Seawolf* in July 1997 reclaimed the title for the United States. The *Seawolf* was designed to be the world's stealthiest submarine, but because of its high cost, the Navy procured only three vessels. The Navy is concentrating now on the new attack submarine to replace the Los Angeles class. The new attack submarine will be designed to operate effectively not only in the blue ocean areas of the sea, but also in green and brown water littoral areas. It will feature low-observability characteristics, such as a modular isolated deck structure, an ultra-quiet propulsion system, sail and hull blending, limber hole covers, and flow control strakes. The design will also take advantage of nonacoustic stealth characteristics, such as an order of magnitude reduction in magnetic signature to avoid magnetic mines, to improve operation in relatively shallow waters. To keep up with the growing use of nonacoustic signatures in antisubmarine warfare, such as wake detection by satellites and improved detection of submerged vessels' infrared and magnetic signatures, the Navy will have to continue to monitor developments in this area.

Both tracked and wheeled tactical vehicles will have built-in low observability features. For example, the future scout and cavalry system will be designed to avoid detection; the Marines' advanced amphibious assault vehicle will have reduced radar, acoustic, and infrared signatures; even self-propelled artillery and tanks, arguably the least stealthy systems, will have reduced visual, acoustic, infrared, and muzzle blast signatures. The design of the Crusader 155mm howitzer and the follow-on to the M1-A2 heavy tank, although still far from stealthy, will also attempt to reduce signatures.

Coatings. Various kinds of coatings have been used on the external surfaces of military vehicles, aircraft, and ships for many years to lessen observability by sensors operating at radar, infrared, and other frequencies. These coatings are generally of three types: appliqué, gel, and liquid. Coatings on new weapons and weapons systems are generally applied in liquid form, and the committee expects this method to continue in the period of interest to this study. Early coatings were applied by hand-held apparatuses operated by highly skilled technicians. This was a reasonable approach when only a few systems had to be treated, although controlling quality and thickness was difficult and the process was expensive. As the demand for coatings grew, robotic and automated application systems were designed and fielded. Various methods of controlling quality and thickness are adequate for coating relatively simple surfaces, but automated application on complex shapes is a difficult engineering feat, with correspondingly high costs. Point of origin process control of the nozzle position and other process variables will have to be improved. Sensors that can operate in the hostile processing environment while the spray is being applied would require a breakthrough in technology and would be valuable for many stealth applications.

Shaping. The shaping of surfaces to deflect radar or acoustic energy away from the emitting source is a well known and commonly used technique. Aircraft and vessels have always been shaped for aerodynamic or hydrodynamic efficiency and performance. However, shaping the surfaces of vessels or aircraft for optimal stealth performance could conflict with aerodynamic or hydrodynamic requirements. According to DTAP, controlling vortex flow and flow separation in low observable configurations is a major technical challenge for air platform technology (DTAP, 1997). DTAP also lists technical challenges for the acoustic signatures of submarines in complex hydrodynamic flows, including improved understanding of hydrodynamic forcing mechanisms and the resulting response and acoustic radiation of structural components, and improved prediction of highly complex hydrodynamic flows to reduce the need for experimental evaluations and to enable the development of propulsors and maneuvering concepts.

Designers of low-observable aircraft and submarines have often been forced to make trade-offs between aerodynamic and hydrodynamic performance and stealth performance characteristics. These trade-offs, in turn, have significantly affected the manufacturability of the vessels. An effective balance was reached for the F-22 Raptor and SSN *Seawolf*, which have good aerodynamic/hydrodynamic characteristics as well as good stealth characteristics. Nevertheless, the development costs, as well as the fabrication and assembly costs, would have been lower if this balancing act had not been required.

The next generation of low-observable aircraft and submarines will require even more difficult design trade-offs, because of projected improvements in the adversary's ability to detect stealth vehicles and the improved performance of potential adversary systems. More affordable manufacturing techniques, processes, and tools that can produce the unusual and complex shapes required for both stealth and aerodynamic/hydrodynamic performance will be necessary, as well as process modeling based on finite-element analysis of materials characteristics during forming.

Gaps and Edges. Manufacturing considerations are often at odds with the requirement for minimal signatures created by weapon structure interfaces. Close horizontal and vertical edge and gap tolerance are necessary to control the reflective surface presented to a radar signal. Another approach is to limit the number of interfaces because systems with few access panels and openings are much less observable to radar. These panels and openings must be designed and manufactured with low observability in mind. A disadvantage of restricting the number of access panels is that it also makes access for maintenance purposes more difficult, time consuming, and, consequently, more expensive. Conformal mold line technology, a relatively new method of covering hinge lines and edges between surfaces, is another approach to closing gaps and covering edges. Like the other new techniques, however, it is expensive and design trade-offs have to be made.

Radar-Absorptive Materials and Structures. New radar-absorptive materials (other than coatings) and radar-absorptive structures could also reduce radar signatures. These materials might have to be load bearing, able to withstand extreme heat (e.g., from jet exhaust), lightweight, formable into very complex shapes with high structural strength, or able to pass some radio frequency signals while preventing the passage of others (e.g., bandpass radomes). These materials and structures are difficult and costly to manufacture.

Shielding. Designers commonly shield hot areas or elements of weapons and weapons systems to reduce their infrared signatures. Sufficient shielding that does not degrade performance or overburden the system with extra weight poses serious challenges for designers and manufacturers.

Sensors

The need for situational awareness and the effective use of weapons requires that most combat systems be equipped with defense-unique sensor systems, e.g., radar, visual aid, forward-looking infrared (FLIR), infrared search and track (IRST), and other multispectral sensor systems.

Radar Sensors. The radar used in military combat systems, such as fighters, bombers, and tanks, is different in purpose and technical requirements from the radar used in commercial aircraft and by law enforcement agencies. Commercial aircraft radar are mainly used for avoiding bad weather and determining the aircraft height above ground in landing approaches. Radar is also used by military aircraft to measure height above ground and help avoid bad weather, but in general, military radar are of the following three types:

- radar designed to find, acquire, and track enemy vehicles, aircraft, or ships and to provide data for missiles or guns used to destroy them
- radar designed to help the crews of combat systems navigate and find, identify, and attack targets on the land or sea surface
- radar designed to enable aircraft to fly "nap-of-the-earth" in terrain-avoidance or terrain-following mode

Radar of any kind on a stealthy combat system poses obvious design and manufacturing problems. Radar dishes or arrays, for example, must be designed to minimize stray and out-of-band emissions. This often requires absorptive coatings and bandpass radomes that allow the transmission of their own energy but shut out externally generated energy. Currently, bandpass radomes are expensive both to design and manufacture.

Infrared Sensors. Many combat systems in service today and planned for the future employ infrared sensors, including FLIR sensors, targeting infrared sensors, and IRST sensors, either installed or carried in pods. These sensors are used for navigation and low-level flight at night and to locate and attack some types of targets. Infrared sensors are not entirely defense-unique because they have limited applications in law enforcement helicopters and light aircraft and are used by agencies involved in border patrol, drug interdiction, and search and rescue operations.

Infrared windows can be made from various materials, all of which are costly. For sensors in high-performance, stealth aircraft like the F-22, both the materials and manufacturing processes are expensive. The military needs less expensive, easier to manufacture, high-performance infrared windows.

Other Electro-optical Systems and Aiming Devices. Gunsights on combat systems are defense-unique but do not pose significant manufacturing challenges. Some combat systems (e.g., the Navy F-14 and some Russian fighters) have telescopes mounted on the wings or fuselage to help identify unknown aircraft at long distances. Although telescope technology is not a defense-unique technology, the application and installation on a combat system can be challenging to both designers and manufacturers (e.g., minimizing aerodynamic drag and signature in fighters).

Self-Protective Sensors. Combat crews and systems are necessarily exposed to enemy defenses in performing their missions. Adversaries around the world are now equipped with sophisticated systems that have high technology sensors and weapons. As 2010 approaches, their equipment will become even more sophisticated and effective.

Aircraft, land vehicles, and ships must be equipped with passive sensors that alert the crew when they are being illuminated by threat radar or other target illuminators and identify the type of illuminator being used, the kind of fire control system associated with it (and thus the type of weapons system it controls), and the direction and approximate range of the emitter. They will also need sensors that can tell them when a missile has been launched against them and its location so they can take countermeasures.

Once the illuminating enemy radar is detected, the combat system crew has several options: evade the enemy system, if feasible; attack the enemy system; or use electronic countermeasures to jam or confuse the system. The latter option is enabled by on-board electronic systems that can receive, analyze, and either jam or deceive the enemy illuminator. Electronic countermeasure equipment has no commercial counterpart. Active protection can be provided by stand-off jammers (e.g., the Navy EA-6B) and by fighters that seek out enemy radar systems and destroy them with defense-unique munitions, such as radar homing missiles. DTAP lists several major technical challenges in the area of warnings against radar threats, including the development of a high-accuracy direction-finding capability; the development of functional elements using monolithic microwave integrated circuits (MMICs); and pulse-level specific emitter identification extraction, processing, and automation (DTAP, 1997).

DTAP also discusses several major technical challenges in the area of missile warning systems, including increasing the detection range of electro-optical/infrared sensors by 100 percent; improving their angle-of-arrival determination to better than one degree; enhancing the probability of detection to more than 95 percent; and reducing false alarms to less than one per hour (DTAP, 1997).

The cost of radomes and infrared windows for these sensors is very high. Therefore, designs for manufacturability and the development of low-cost production processes will be necessary. Low-cost designs and processing will also be necessary for the next generation IRST sensors, as well as the development of packaging for functional elements using MMIC, which promises to reduce electronics to one-third of their current volume.

Electronics

Semiconductor electronics and photonics are critical for avionics, communications, surveillance, control, and other military applications. By 2010, they will be ubiquitous in all elements of defense systems from sophisticated space platforms and sensors to communications and vision systems carried by individuals. They will have to operate in a wide range of temperatures, extremes of humidity, high radiation, and other hostile environments.

Military applications will require that more and more high-performance transistors and lasers be packaged at a higher density while simultaneously improving the quality and reliability of the system. Consistent with industry trends, system-on-a-chip architecture is expected to play an increasing role, as chip-level integration is used in lieu of system packaging to achieve the desired functional density. Even though the use of electronic devices in the engine compartments of automobiles is driving improvements in technology performance in hostile engine environments, certain military systems will expand the performance envelope significantly beyond evolving commercial capabilities.

The annual cost of avionics maintenance in the four services is staggering. The Rand Corporation recently estimated that the annual cost of avionics maintenance in the F-16 system alone was more than $100 million (Stevens et al., 1997). The cost of avionics is steadily increasing in all defense systems, with software maintenance and nondigital functions being the main drivers of life-cycle costs. The corrosion and fatigue/structural failure of connectors are prominent problems. Past studies by the ManTech program have shown that fatigue, corrosion, and thermal cycling failures are critical factors that must be addressed in the specifications of a system (ManTech, 1998).

Avionics software management budgets are routinely inadequate. For instance, the projected cost of required F-15 development and flight testing for fiscal years 1997 through 2002 is approximately $500 million; the shortfall in the projected budget is about $140 million. If emulators and automated validation tools could be used to replace flight tests and proprietary interfaces and technology could be eliminated by using existing commercial open systems, software management costs in many systems could be reduced significantly.

Periodic updates or modifications of avionics systems are often hindered by the high cost of rewiring older systems. Bridging existing networks by means of field programmable gate arrays, with new wiring and commercial protocols, could ameliorate this problem.

Avionics maintenance costs in older aircraft often consume the funds available for periodic updates. The following measures could be taken to correct this situation:

- improve packaging to increase structural reliability and reduce connector problems
- improve built-in test diagnostics to reduce "retest OKs" and reduce the amount spent on automated external test equipment
- use modular/throwaway components to facilitate maintenance by eliminating the need to return the components to a depot and repair them
- develop prognostic capabilities, or intelligent system health monitors, to facilitate maintenance and reduce life-cycle costs
- replace military specification cards with COTS hardware to lower costs substantially and improve reliability (provided that the hardware can withstand the required environmental stresses or can be mounted on shock mountings or otherwise protected)
- replace or interface, where feasible, existing buses and networks with commercial programmable network protocols to reduce costs

- develop and test/demonstrate software reengineering tools to facilitate upgrades to cope with the rapid obsolescence of electronic technology

Closing the gap between the needs of defense system electronics and commercial developments will require specific technological improvements, such as lightweight chip-on-board (also called flip-chip or direct chip attachment) platforms that feature electronic miniaturization. These platforms reduce board area by as much as 50 percent and component weight by as much as 80 percent over packaged devices; the thermal load, however, is increased dramatically. Widespread commercial applications of this technology for single-chip packaging, multichip packaging, and direct attachment to printed wiring boards are expected by 2010. For advanced military applications, reliability measures will include thermal shock resistance, thermal cycling fatigue, temperature or humidity bias, and mechanical shock and vibration resistance. In addition, new materials and processes may be needed for use in harsh environments.

Failures in electronic interconnects are of ongoing concern in defense applications. Fatigue, corrosion, and wear contribute to both short-term and long-term failures. As data rates and bandwidth increase, the manufacture of high-precision, high-reliability connectors, back planes, and traces will confront physical barriers. Connector fretting studies, plated-through-hole thermal fatigue studies, studies related to dendritic growth phenomena in fine-pitch devices, studies of chip-on-board packaging technology, and other studies and experimentation will be necessary to ensure reliability.

High-speed electronics are susceptible to failure from microsecond interruptions. As the speed of electronics increases, interruption-free connector systems will have to be designed and connector integrity will have to be quantified in severe vibration environments. Optical interconnections will be essential to meeting the challenge of ultra-high data rates. Improved fiber-optic connectors and wiring capable of functioning reliably in severe vibration environments should be investigated. As the thickness of multilayer boards continues to increase, plated-through-hole thermal fatigue will become even more of a problem than it is today. Military applications for multilayer boards are numerous and will continue to increase. The Japanese electronics industry has addressed the problem at the first indenture by improving the ductility of copper used in plated-through-hole applications, but much more remains to be done in both in the laboratory and to improve manufacturing processes.

As electronic devices become more densely packaged, the fine-pitch aspects of the designs become more susceptible to dendritic growth, which results in intermittent failures. Studies could address the limits of fine-pitch capabilities in humid environments with thermal cycling and power cycling. Various conformal coating techniques and capacities could be investigated and documented, if not improved.

The United States, which has no significant commercial liquid crystal display

(LCD) manufacturing industry and a very limited military LCD industry, depends on foreign sources for LCD technology. No growth in U.S. commercial capabilities is expected by 2010, and economic pressures are expected to continue to erode the military base. Existing foreign commercial LCD technology cannot satisfy future military requirements for display panel and electronic interconnections. Areas that will require study, documentation, development, and testing include: basic glass manufacturing technology; brightness, dynamic range, and viewing angle; mechanical shock and vibration resistance; and thermal cycling fatigue and temperature/humidity. In addition, competing technologies, such as electroluminescent and plasma technologies, require study.

Information Systems

Applications for information technology are pervasive in defense operations and weapons systems. However, this section will deal only with the capabilities of information technology for defense manufacturing. Based on DTAP, defense technology objectives, and Mantech planning documents, the capabilities of information technology required for defense manufacturing fall into three categories: interoperability with commercial systems; information requirements for defense-specific products; and information security. Some of the specific capabilities are listed below:

- systems architectures that permit the secure use of COTS computers, software, and networks
- interoperability of defense logistics systems and the diverse systems used by suppliers
- network management and control protocols for data security in distributed design and manufacturing operations to prevent interruption, jamming, sabotage, and interception
- models for defense products with multiple levels of resolution to enable simulation-based design
- databases of weapons system life-cycle costs that can be integrated into design systems to enable life-cycle cost trade-offs simultaneously with design evolution
- production process capabilities and cost databases that can be integrated into design systems to provide simultaneous assessments of design alternatives and production costs, manufacturing risks, and manufacturing systems designs
- product data models and storage and retrieval architecture capable of seamlessly handling all data modalities
- product structure directories to meet unique structural requirements for defense products that also have open architecture and meet commercial standards

- mechanisms, including intelligent agents, for locating and retrieving information from complex database structures
- automated systems for reverse engineering based on scanning of an actual part
- parametric modeling to enable design trade-offs at the conceptual level

Manufacturing Processes and Technologies

Production Rate Transparency

Because of declining defense budgets and the resultant reductions in new weapons systems, few, if any, major weapons systems will be produced at a high volume in the foreseeable future. This is not to say that certain defense items, such as munitions, will not be produced in high volume. But the production of complex systems will be characterized by very low throughput, which raises the question of whether industry is prepared for the economical production of goods and systems at very low rates.

Experience has shown that unit costs increase significantly as production rates drop. The committee believes this problem could be ameliorated by focusing attention on manufacturing technology for low-rate production. Ideally, the production of defense goods will be "rate transparent," i.e., a component, subsystem, or complex weapons system will be produced at the same cost regardless of the production rate. Manufacturing capabilities that will be critical to minimizing unit production cost at low production rates include: flexible production lines, procurement of materials in bulk, modeling of production during the design process, and adaptive process control to achieve 100 percent first time yields.

Repair of Parts Made of Composite Materials

Currently, repairs of parts made of composite materials require a high degree of operator skill and long cycle times. An automated process could make machine-generated scarf cuts. Because much of the patch bonding will be done on aircraft, ships, vehicles, or other systems with composite skins and structures, efficient and affordable technologies and processes will be required for on-system, on-site repairs, as well as depot repairs of damaged composite structures and surfaces.

Dimensional Control

The need for close gap tolerances in systems requiring stealth was discussed earlier. But many other areas of defense manufacturing also require dimensional control and tight tolerances. For instance, submarine construction requires circularity and hull fairness, as well as close control of tolerance stack-ups to facilitate modular construction. Interestingly, the same techniques are now being used to achieve low observability for the surface fleet.

Controlling distortion, hardware variability, and dimensional accuracy can minimize dimensional variations and ensure the efficient manufacture and assembly of parts. The goal is to achieve a highly capable fabrication/assembly/construction process that consistently meets required specifications without reworking. Ideally, dimensional control will begin in the early stages of design when construction process capabilities and design specifications are being assessed for compatibility.

Tight tolerances are required for many defense systems, such as those produced in modules (e.g., modern jet engines, submarines, ships, avionics, aircraft, and land vehicles). Modular construction requires more stringent fit and tolerance control than nonmodular construction. Manufacturing process capabilities and assembly sequences must be clearly and accurately defined to determine the dimensional tolerance stack-ups associated with efficient modular construction. Tolerance stack-ups at interfaces between modules or assembled parts must be planned for early in the design phase by design details that accommodate expected variations.

As tolerances are tightened, manufacturing becomes more difficult, rejection and reworking increase, and costs go up. Experienced designers and production engineers always try to allow for the loosest tolerance for the end product to function effectively. Many modern defense systems, however, such as stealth aircraft, submarines, and other systems, require tolerances that were not even considered achievable a few years ago without significant added expense and manufacturing process time.

Modern tooling and processes have significantly reduced the cost of maintaining very tight tolerances in fabrication and assembly. The design part geometry from computer-aided design and manufacturing (CAD/CAM) is used in a variety of manufacturing applications related to dimensional control. This geometry can be expanded to account for shrinkage due to welding and incorporated into the numerically controlled code for the automated marking, forming, cutting, fitting, and welding of parts. The electronically developed geometry representing the ideal condition at different stages of fabrication, construction, and assembly can be used for comparison with as-built part dimensions. Industrial measurement systems, such as photogrammetry, multitheodolite laser trackers, and total stations, which provide highly accurate electronic capture of as-built parts, can provide a statistical definition of process capability. This capability can be used as a scientific basis for improving processes and reducing rework. The application of advanced computer-aided visualization techniques can provide a thorough understanding of dimensional changes throughout the construction process and help determine specific process changes to improve dimensional quality.

The key term in dimensional control is "as-built," which assumes post-operation (e.g., after the part has been machined) inspection. This approach usually results in some degree of scrapping, reworking, and repair. To minimize these, noncontact inspection during the operation will be necessary.

The committee was informed by several industry and laboratory sources (West, 1998) that little is currently being done to integrate, in an automated way, product analysis and design with tool design and manufacturing processes. The key to improving dimensional quality is the systematic identification and control of process variables. The data sampling and analysis needed for constant monitoring of manufacturing processes and continuous improvement of quality must be done in an integrated way. Process data systems will be required that can capture and transmit data between manufacturing processes and design and analysis systems in an integrated way, cost-effectively, accurately, and quickly. Data systems should fully integrate all facets of product analysis and design, manufacturing process analysis and design, tool analysis and design, and inspection/control system analysis and design.

CAD systems should be capable of automating the expansion of part geometry and associated attributes, such as layout and reference lines, to account for weld shrinkage. In addition, these systems should be able to transfer numerical control data to automated marking, cutting, fitting, and welding processes; efficiently transfer geometric configuration data to and from industrial measurement systems; and provide efficient analysis and visualization of comparative data between ideal and as-built products.

Automated manufacturing processes and inspection and measurement systems should provide highly accurate and automated dimensional quality control in making, cutting, forming, assembling, and welding parts using tools and processes, such as lasers, water jets, electron beams, and high-speed machining. Automated, highly accurate systems are needed for verifying the accuracy of assembly tools and component locations. Today's labor-intensive methods require inspectors, tool templates, and gauges. A new verification method should use advanced photographic or laser technology.

Titanium Processes

Problems are caused in titanium investment casting when small pieces of the ceramic face material used on the inside of the investment casting mold break loose and migrate into the molten titanium. There is currently no nondestructive inspection (NDI) method capable of reliably detecting fusion defects, ceramic shell inclusions, and regions of dissolved shell in titanium castings. These problems are particularly burdensome in certain modern aircraft, which have significant weight constraints and require large titanium castings for high strength. Improved NDI methods are especially important for the F-22 program, which requires two 5-foot-long (1.5 m), 200-pound (90.7 kg) castings as part of the wing attachment fittings. Currently, extensive radiographic inspection is necessary to ensure that castings are free of these defects. Even after extensive inspections after casting, defects are sometimes detected after machining, which has significant cost implications.

A method of detecting defects early in the casting cycle would allow a decision to be made to rework a part prior to heat treatment or to discard the part without incurring additional costs; this might also help to identify the cause of the defects and lead to process improvements. An effective NDI method would have to reveal defects for complex geometries and thickness up to three inches with more than 90 percent reliability. In addition, the method would have to have a minimal effect on casting times and cost.

At present, no robust coating for large structural titanium investment castings exists that produces limited reaction with molten titanium and is readily detectable by available NDI techniques. Such a coating would significantly reduce shell inclusions and make it easier to detect inclusions when they occur. In addition, reducing the reaction with the shell material would improve the quality of the casting surface and reduce the need for reworking surfaces. If titanium honeycomb could be produced from alloy 15-3, it would provide a much better strength-to-weight ratio, would not be subject to node failures, and would be an order of magnitude less expensive than graphite composite core.

Overall Process Optimization above the Plant Floor

Optimal effectiveness and efficiency of manufacturing systems will require improvements "above the plant floor," as well as improvements on the floor. The establishment of a nonrecurring manufacturing process control requires simultaneous product and process views, single view management, a single numbering system (e.g., for work orders, work breakdown structure, shop orders, drawing numbers, part/assembly numbers), a visual statusing system, visibility of upstream problems and downstream impacts, and drill-down expansions and database linkages.

SUMMARY

Required manufacturing capabilities, based primarily on DTAP, are summarized in Table 2-1. The committee analyzed these and determined that they fall into six broad technology categories: composite processing and repair; electronics processes; information technology systems; sustainment; design, modeling, and simulation; and production processes. Table 2-2 lists the manufacturing capabilities that fall under each of these categories. Some of the required manufacturing capabilities identified by the committee are specific to certain weapons systems (e.g., processes for radomes and infrared windows and processes for munition fuzes). Others are applicable to a number of weapons systems. Widely applicable capabilities are listed below:

- simulation-based design (including product and process models) able to make cost versus performance trade-offs during design and simultaneously design products and their manufacturing processes
- cost versus performance trade-offs at the conceptual level of the design process
- product data structures that meet the unique characteristics of defense products
- interoperability of defense information systems and commercial systems
- low-cost composite structures through novel designs and new processing concepts
- new system and component design concepts to enable electronics (including COTS products) to operate reliably in harsh military environments
- open-system architectures (including modular designs) to facilitate upgrading systems and accommodate unexpected changes in the availability of parts
- intelligent health monitoring systems for electronic mechanical subsystems with predictive capabilities to facilitate maintenance
- dimensional control in large structures
- adaptive process controls to improve first-time yields
- life-cycle cost analyses concurrent with design
- low-rate production methods

TABLE 2-1 Required Defense Manufacturing Capabilites Based on the Defense Technology Area Plan

Technology Area	Manufacturing Capability
Weapons System Platform Technologies	
Aircraft weapons systems	Repair techniques for aging systems
	Nonintrusive, real-time monitoring techniques for flight loads and damage
	Design techniques and processing methods for high strength-to-weight materials, particularly composites
	Design concepts and processing methods that reduce the costs of composite structures
	Electronic systems able to withstand high g loads and severe vibrational environments
	Affordable processing methods for launch equipment with reduced drag and signature
	Weapons systems capable of launching weapons at high speeds and under high g loadings
Surface and subsurface sea combat vessels	Design concepts that minimize weight and volume of vessel systems and reduce life-cycle costs
	Automated, intelligent monitoring and control systems
	System-level design approaches to reduce acoustic signatures and cost, and increase shock resistance
	Design simulations to enable accurate performance versus cost trade-offs
Land combat vehicles	Maintenance and upgrade technologies for aging systems
	Integrated product and process development
	Virtual prototyping
Weapons Technologies	
Expendable munitions	High-yield, robust fuze production process
	Methods for precise filling of explosives in munitions
	Automated filling of explosives in munitions to increase safety, improve process yield, and ensure performance

continued

TABLE 2-1 continued

Technology Area	Manufacturing Capability
Missiles and torpedoes	Methods for miniaturizing system components Low-cost production processes Composite materials for advanced propulsion systems Methods to reduce cycle time and nonrecurring costs in production processes Overall system designs based on common subsystems
Guns	Methods to reduce cycle time and nonrecurring costs
Mobile weapons systems	Methods for packaging electrothermal chemical technology Designs for high-efficiency plasma ignitors and high-energy-density propellants Designs for high-efficiency rails Designs to minimize weight and size of components
Cross-cutting Technologies	
Low observability technology	Precise, automated methods for applying low observability coatings Process control sensors that can operate in hostile processing environments Affordable manufacturing techniques, processes, and tools that can form complex shapes with high stealth and aerodynamic/hydrodynamic performance Process models based on finite-element analysis of materials characteristics during forming Conformal mold line technology Methods for design trade-offs to minimize signatures created by gaps and edges Radar-absorptive materials and structures that are strong, lightweight, able to withstand extreme heat, formable into complex shapes, and affordable Designs for lightweight, effective infrared shielding
Sensors	Designs for high-performance radomes and infrared windows that are affordable and easy to manufacture Designs for electro-optical systems that are affordable, easy to install, and that have minimal drag and signatures High-density packaging for functional elements using monolithic microwave integrated circuits

TABLE 2-1 continued

Technology Area	Manufacturing Capability
Electronics	Automated validation tools to replace flight testing
	Commercial software systems to replace proprietary systems
	Methods to bridge existing networks using field programmable gate arrays, new wiring, and commercial protocols
	Avionics packaging with increased structural reliability and reduced connector problems for aging systems
	Built-in test diagnostics for aging systems
	Modular components to facilitate maintenance of aging systems
	Intelligent health monitors for aging systems
	Commercial hardware to replace military specification cards and improve reliability
	Commercial programmable network protocols to replace existing buses and networks and reduce costs
	Software engineering tools to facilitate upgrades and cope with rapid obsolescence of electronic technology
	Lightweight chip-on-board platforms that feature electronic miniaturization
	Platforms with reliability in terms of thermal shock resistance, thermal cycling fatigue, temperature and humidity tolerance, and mechanical shock and vibration resistance.
	Materials, components, and processes that can be used in harsh military environments
	High-precision, high-reliability connectors, back planes, and traces
	Interruption-free connector systems
	Optical interconnections for ultra-high data rates
	Manufacturing processes for multilayer boards
	Conformal coating techniques and capacities to prevent dendritic growth
	Glass manufacturing technology for liquid crystal displays

continued

TABLE 2-1 continued

Technology Area	Manufacturing Capability
Information systems	Systems architecture that permits secure use of commercial-off-the-shelf computers, software, and networks
	Defense logistics systems that are interoperable with the diverse systems used by suppliers
	Network management and control protocols to ensure data security in distributed design and manufacturing operations
	Product models with multiple levels of resolution for simulation-based design
	Databases containing weapons system life-cycle costs for integration into design systems
	Production process capabilities and cost databases for integration into design systems
	Product data models and storage and retrieval architectures capable of handling data seamlessly
	Product structure directories that are open and meet commercial standards
	Intelligent agents for locating and retrieving information
	Automated reverse-engineering systems based on scanning of the actual part
	Parametric modeling to enable design trade-offs

Manufacturing Processes and Technologies

Production rate transparency	Flexible production line
	Procurement of materials in bulk
	Methods for modeling production processes during design
	Adaptive process controls to enable 100 percent first time yields
Composite repairs	Automated composite repairs
	On-system, on-site repair technologies and processes that are affordable and efficient

TABLE 2-1 continued

Technology Area	Manufacturing Capability
Dimensional control	Manufacturing processes and assembly sequences that determine dimensional tolerance stack-ups for modular construction
	Design methods that incorporate tolerance stack-ups at interfaces between modules or assembled parts
	Measurement systems that provide highly accurate electronic information on as-built parts
	Computer-aided visualization techniques
	Noncontact inspection during manufacturing operations
	Process data systems that integrate product analysis and design, manufacturing process analysis and design, tool analysis and design, and inspection/control system analysis and design
	Computer-aided design systems that integrate design, production processes, measurement processes, and compare ideal and as-built products
	Automated, highly accurate dimensional control systems using advanced photographic or laser technology
Titanium processes	Nondestructive inspection technology for titanium castings
	Method for coating structural titanium investment castings that produces limited reaction with molten titanium and where inclusions are detectable
	Process for producing titanium honeycomb from alloy 15-3
Overall process optimization above the plant floor	Nonrecurring manufacturing process control with single view management, single numbering system, visual statusing system

TABLE 2-2 Broad Categories of Required Defense Manufacturing Capabilities

Category	Manufacturing Capability
Composites processing and repair	Design methods and processes for low-cost structural composites
	Design methods for low-cost composite materials
	Composite materials for advanced propulsion systems
	Low-cost composite surfaces for tactical missiles
	Automated composite repairs
	On-system, on-site composite repair technologies that are affordable and efficient
Electronics processes	Intelligent health monitoring systems
	Electronic systems able to withstand high g loads and severe vibrational environments
	High-density packaging for functional elements using monolithic microwave integrated circuits
	Electronics packaging with increased structural reliability
	Built-in test diagnostics
	Commercial programmable network protocols to replace existing buses and networks
	Software engineering tools to facilitate upgrades
	Lightweight chip-on-board technology for miniaturization
	High-precision, high-reliability connectors, back planes, and traces
	Interruption-free connector systems
	Optical interconnections for ultra-high data rates
	Designs to prevent dendritic growth in high-density electronics
	Manufacturing technology for liquid crystal displays
Information technology systems	Commercial software systems to replace proprietary systems
	Systems architecture that permits secure use of commercial off-the-shelf computers, software, and networks
	Defense logistics systems that are interoperable with the diverse systems used by suppliers
	Network management and control protocols to ensure data security in distributed design and manufacturing operations
	Databases containing weapons systems life-cycle costs for integration into design systems
	Production process capabilities and cost databases for integration into design systems

TABLE 2-2 continued

Category	Manufacturing Capability
Information technology systems *continued*	Product data models and storage and retrieval architectures capable of handling data seamlessly Product structure directories that are open and meet commercial standards Intelligent agents for locating and retrieving information Automated reverse-engineering systems based on scanning of the actual part Nonrecurring manufacturing process control with single view management, single numbering system, and visual statusing system
Sustainment	Repair techniques for aging systems Nonintrusive, real-time monitoring techniques for flight loads and damage Maintenance and upgrade technologies for aging systems Automated validation tools to replace flight testing Avionics packaging with increased structural reliability and reduced connector problems for aging systems Built-in-test diagnostics for aging systems Modular components to facilitate maintenance of aging systems Software engineering tools to facilitate upgrades
Design, modeling, and simulation	Product models that enable accurate life-cycle performance versus cost trade-offs Integrated product and process development Virtual prototyping System designs based on common subsystems Process simulations based on finite-element analysis of materials characteristics during forming Product models that enable stealth versus other performance characteristics trade-offs Designs for affordable, high-performance radomes and infrared windows Designs for affordable, easy-to-install electro-optical systems with minimum drag and signature Product models with multiple levels of resolution to enable simulation-based designs Parametric modeling to enable design trade-offs

continued

TABLE 2-2 continued

Category	Manufacturing Capability
	Integrated product, tool, and manufacturing process designs
	Design methods that incorporate tolerance stack-ups
	Computer-aided design systems that integrate design, production processes, measurement processes
Production processes	Affordable processing methods for launch equipment with reduced drag and signature
	High-yield, robust fuze production process
	Methods for precise filling of explosives in munitions
	Automated filling of explosives in munitions to increase safety, improve process yield, and ensure performance
	Methods to reduce cycle time and nonrecurring costs in production processes
	Precise, automated methods for applying low observability coatings
	Affordable manufacturing techniques, processes, and tools that can form complex shapes
	Conformal mold line technology
	Manufacturing processes for multilayer boards
	Conformal coating techniques to prevent dendritic growth
	Glass manufacturing technology for liquid crystal displays
	Flexible production lines
	Adaptive process controls to enable 100 percent first-time yields
	Manufacturing processes and assembly sequences that determine tolerance stack-ups for modular construction
	Measurement systems that provide highly accurate electronic information on as-built parts
	Computer-aided visualization techniques
	Noncontact inspection during manufacturing operations
	Automated system for accurate location of assembly tools and components
	Nondestructive inspection for inclusions in titanium castings
	Process for producing titanium 15-3 honeycomb

3

Leveraging Advances in Commercial Manufacturing

INTRODUCTION

Manufacturing technologies are advancing rapidly, and, by 2010, new technologies and practices will lead to even more dramatic improvements in quality and productivity. To meet the challenges described in Chapter 1 and develop the required defense capabilities described in Chapter 2, defense manufacturing must take advantage of technology advances being pioneered by the commercial sector in areas applicable to defense products. A growing trend among government and defense manufacturers is the adoption of commercial "best practices." In addition, many companies have combined commercial and defense manufacturing processes and products to take advantage of economies of scale in facilities, resources, and organizational structure. This approach also provides opportunities for leveraging suppliers, material purchases, and systems. Many defense contractors are working to integrate their defense and commercial manufacturing operations, and this trend will continue.

Commercial manufacturers have become increasingly important as sources of defense products. The increased emphasis on low cost has led DOD to promote the use of COTS products and to investigate the possible manufacture of defense products on commercial production lines. Because commercial industry is much larger than the defense industry, it has a correspondingly stronger base for technology development and manufacturing advances. For example, the commercial electronics industry will provide most of the advances in technology and manufacturing for defense products (with the exception of some defense-unique products). The advantages of COTS hardware and software include much lower development costs, tested reliability and performance, and substantially shorter product cycles.

ADVANCES IN COMMERCIAL MANUFACTURING

To evaluate advances in commercial manufacturing and identify those with potential for meeting the needs of defense manufacturing, the committee reviewed forward-looking manufacturing studies, including Next Generation Manufacturing (NGM, 1997) and Visionary Manufacturing Challenges for 2020 (NRC, 1998). In addition, the committee reviewed information sources available on the World Wide Web (see Appendix B) and invited speakers to assess advances anticipated in manufacturing. The committee then identified the following areas of management and technology advances that defense manufacturing can expect to draw on:

- industry collaboration
- adaptive enterprises
- high-performance organizations
- life-cycle perspectives
- advanced manufacturing processing technology
- environmentally compatible manufacturing
- shared information environments

These advances are interactive, rather than being independent of each other (i.e., shared information environments may support the goals of high-performance organizations and adaptive enterprises may promote industry collaboration). Each of these advances is composed of a number of elements, which may also be applicable to other areas. These elements are summarized in Table 3-1 and described below.

Advanced Approaches to Manufacturing Accounting

Activity-Based Accounting

In activity-based, or process-based, accounting, costs are assigned to the actual activity or process in which they occur. In conventional systems, average costs are allocated per product. Total product costs are the same in both approaches, but, in the conventional system, it is impossible to assess cost drivers. The disadvantage of changing to activity-based accounting is the effort involved in revamping existing systems.

TABLE 3-1 Commercial Manufacturing Advances and Elements

Advance	Elements
Industry collaboration	Electronic commerce Teaming among organizations Long-term supplier relationships Virtual enterprises
Adaptive enterprises	Agile enterprises Reduced lead time Reduced cycle time Activity-based accounting Lean enterprises Knowledge-based and learning enterprises
High-performance organizations	Virtual co-location of people High performance work teams Cross-functional teams
Life-cycle perspective	Standardization of parts and reduction in number of parts Integrated product and process development Life-cycle design Cost as an independent variable accounting
Advanced manufacturing technology	Flexible assembly Soft tooling Single piece fabrication Rapid prototyping Three-dimensional digital product models High-speed machining Simulation and modeling Predictive process control technologies Adaptive machine control Tool-less assembly Nanotechnology Biotechnology Embedded sensors Generative numerical control Flip chips
Environmentally compatible manufacturing	Cleaning systems Coating systems Material selection, storage, and disposal
Shared information environment	Data interchange standards Internet, intranets, and browser technology Intelligent agents Seamless data environment Telecommunications Distance learning

Accounting with Cost as an Independent Variable

When cost is considered as an independent variable (CAIV) in accounting, it can be treated as a fixed parameter, and performance and design criteria can be traded off to meet cost objectives. This cost accounting method can be used to improve design decisions.

Advanced Approaches to Product Design

Life-Cycle Design

The life-cycle perspective takes into account the entire life cycle of a product; thus life-cycle design includes designing for all stages (initial development through disposal) and all aspects of a product's producibility, reliability, maintainability, and affordability. For example, life-cycle design can include "design for assembly," i.e., design of parts aimed at decreasing the time and cost required for product assembly. The life-cycle design process incorporates product disposal considerations by selecting materials that are recyclable or easily disposable. In some cases, materials decisions have to take into account environmental legislation, as is the case with the use of fluorocarbon refrigerants and radioactive components. The cost of virgin materials versus reclaimed materials can also be considered.

Studies have shown that more than 80 percent of product costs are established during the design process. Life-cycle design can, therefore, have a major impact on total life-cycle cost, especially for long-lived products. Sophisticated cost analysis, such as CAIV, and design trade-off tools, as well as more open communication, linkages with supplier capabilities and costs, and interactive iterative dialogues with customers will be required.

Integrated Product and Process Development

Integrated product and process development (IPPD) emphasizes the timely collaboration of stakeholders, including customers and key suppliers, in a systematic development process. IPPD encompasses new business development, research and development, product and process development, transition to production, and continuous product improvement. IPPD is a standard framework and culture for operating a design-engineering, manufacturing, support enterprise that integrates the customer, sites, and suppliers. Products are designed for manufacturability, assembly, and support. A number of different approaches to IPPD can be taken, the success and appropriateness of which depend on the nature of the product, the culture of the organization, and the team members involved. IPPD can be used to incorporate life-cycle perspectives, make cost versus performance trade-offs, and run simulation models to evaluate design alternatives.

Standardization and reducing the number of parts is a demonstrated approach to reducing costs in components and assemblies. Using commercially available components and standard parts wherever possible can reduce costs. Modular designs allow for more flexibility in technology upgrades and component replacements. In addition, organizations can apply standards to the processes involved in designing, producing, and supporting products, although they must also relinquish some creative freedom.

Three-Dimensional Digital Product Models

A three-dimensional digital model of a product fully describes the geometry of the product, materials to be used, the attributes of its parts, and the relationship between its parts. In other words, such a model includes all critical information regarding a product's physical dimensions. CAD packages, increasing computer power, and robust product data management systems have enabled this technology, which can manage, in a configuration-controlled environment, almost all descriptive data about a product and can be used to provide an electronic mock-up (in place of a physical mock-up) during product development. The flow of data between the phases of product development must be seamless, thereby eliminating the revision or reloading of data with each new phase. Digital models must be able to automatically generate the preferred manufacturing plan or process. Product model data can be used to provide real-time manufacturing cost estimates. When electronic "prototypes" replace physical prototypes, the need for some physical testing to confirm the design and performance may be eliminated.

Visualization, two-dimensional and three-dimensional representations of objects based on digital source data (e.g., an electronic mock-up of a product under development), is increasingly being used as an extension of CAD. Design data is normally viewed through the CAD platform and requires high-end graphics facilities to display midweight and heavyweight models. Group visualization sessions are used by product development teams to facilitate understanding by all stakeholders of the product being developed. Visualization facilities today are stationary and relatively expensive to build and outfit.

Distributed visualization will be possible when two-dimensional and three-dimensional information becomes accessible through low-cost, standards-based, decentralized viewing devices. Visualization aids at job sites can show manufacturing sequences and assist with work operations, such as the assembly of complex items. Nongraphical data can be appended or incorporated into product visualizations to convey a richer understanding of the product. The in-process status of a complex assembly might be shown by using different colors for installed and uninstalled pieces.

Simulation and Modeling

Simulation-based design utilizes the product model data to simulate processes and events that will occur during the product life cycle. Simulation and/or modeling techniques can predict outcomes in product development, such as product performance, fabrication processes or equipment, software checkout, product testing, and product flow. By replication, these processes can be adapted as necessary before a commitment is made to a prototype or production mode. Techniques such as variation simulation analysis (VSA) and factory floor layout simulation can improve product performance. Assembly modeling can be used to complement simulations to determine if changing the order of steps in the assembly of a complex product can lead to labor savings and reduce variation. CAIV accounting can be facilitated by modeling to determine if design trade-offs will reduce costs or improve performance. Combining three-dimensional product modeling with simulation techniques can help determine the cost of alternative manufacturing processes.

Rapid Prototyping

The development of a physical product includes design, materials planning, process planning, and physical manufacturing. Rapid prototyping can reduce lead time by creating a physical likeness of a product directly from a three dimensional model. Rapid prototyping of single detail parts or one-piece models of subassemblies is typically accomplished using stereolithography, which provides exact physical likenesses of products fabricated from specialized polymers. The prototypes are accurate in physical dimensions and shape, but do not allow for testing of material properties of the production material. Work is being done on using prototypes to create molds for castings. Technically complex products could be built in quicker development cycles and at lower cost if soft tooling, three-dimensional digital product data, and generative numerical control were used. Prototyping of individual components by stereo lithography could be complimented by the fabrication of components from their actual production material through the use of emerging technologies, such as three-dimensional printing, which uses metallic particles deposited layer by layer and a binding process to fuse these layers to create a metallic prototype component.

The development time of software products is being dramatically reduced by means of rapid prototyping tools that provide the look and feel of the end product in less time than conventional methods. Reusable software components, called "objects," are created and custom assembled to perform desired functions. Standards for objects and visual programming tools can support the rapid assembly of objects and other reusable components into prototypes or useful products that can then be used for proof of concept by the end user. Libraries are being developed for objects, specialized components are being created, and object library techniques are being improved so that software developers can readily find needed components.

Advanced Approaches to Manufacturing Processes

Generative Numerical Control

Generative numerical control (GNC) is the automatic creation of numerically controlled programs for numerical control equipment as the designer creates the three-dimensional product data set. Automated numerical control is a stepping stone towards GNC and requires that parts be designed using removal volumes (i.e., removing sections of material to arrive at the finished part, similar to the way a machinist creates the part). Once removal volumes have been established, they can be used as subroutines for creating the numerical control program.

GNC on the factory floor can be coupled with other knowledge bases to reduce flow times and configured to automatically generate the manufacturing plan or process concurrently from the three-dimensional data set. GNC will be able to generate the numerical control program to coordinate measuring machines for quality assurance. Cost figures can be tied to removal volumes, so that engineers have real-time cost visibility of parts as the design is being developed.

Adaptive Machine Control

Adaptive machine control is the ability, in real time, to monitor a process insitu and automatically adjust the process to eliminate variations. Statistical process control uses data collected in real time and charted by operators, but can only use measurable data. New sensors will be necessary to collect several process parameters that would alert the operator to process variations. For example, machine tool spindle speed and force can be measured, as well as cutter location, but real-time data on the actual amount of material being removed at the cutter tip cannot be measured. This measurement would tell whether the desired product dimensions were being generated and would allow immediate control feedback to prevent variations before they occur. This data could be collected with sensors, transducers, and softeners. Using sensors coupled with three-dimensional data sets, products could be inspected in real time against dimensional properties.

Predictive Process Control Technologies

Predictive process controls can analyze and predict variations in a fabrication process, enabling the process configuration (e.g., process steps and equipment) with the least variation and a reasonable cost to be found. VSA is one example of predictive process control. In the automotive industry, key characteristics (the measurable qualities of a part) that directly affect customer form, fit, or function requirements are identified and VSA utilized to determine the process that will best deliver these key characteristics. VSA and other predictive process control methods could also be used on the factory floor to assist tool designers, process planners, and process engineers in implementing process changes or procedures to achieve real-time reductions in variations. These capabilities could allow a reduction in statistical process control data collection by ensuring that the process, not just the product, is robust.

High-Speed Machining

In high-speed machining (HSM), the cutting tool spindle operates at speeds of more than 30,000 rpm and feed rates of more than 200 inches per minute (508 cm/min). Conventional machining parameters are spindle speeds of less than 10,000 rpm and feed rates of less than 100 inches per minute (254 cm/min). HSM depends on the rapid removal of metal chips so that heat generated during the machining process is not transferred to the part being machined. Aluminum and other soft metals can be machined using HSM. Parts with extremely thin final cross-sectional thickness can be machined from billets, effectively replacing sheet metal fabrications. With hard metals, such as titanium, the material may weld itself to the cutting tool. In the future, equipment and cutting tools must be made more reliable; numerically controlled equipment with increased reliability and tolerance control will be important, as will cutting tool technologies that increase metal removal rates. Parts could be designed to take advantage of the reduced flow time and the use of single-piece fabrication with HSM.

Flexible Tooling and Soft Tooling

Flexible tooling is tooling that can be used to assemble more than one product and can thereby reduce nonrecurring costs by eliminating the cost of dedicated tooling. Soft tooling is tooling constructed from nontraditional materials (e.g., wood or foam), instead of the traditional materials used for hard tooling (e.g., metal). Soft tooling has advantages for rapid prototyping, where tooling must be built quickly and at low cost. The disadvantages of using soft tooling include its inability to withstand autoclave processes and concerns about configuration control at high production volumes.

Tool-less Assembly

Tool-less assembly, or determinate assembly, is the joining of detail parts to form subassemblies or the joining of subassemblies to form final products without using tooling or locating fixtures. Tool-less assembly is accomplished by either predrilling or machining parts so that all parts are self-locating or by making a few critical locating points on one (primary) part and installing other parts relative to the primary part. In this scenario, a part can be thought of as a tool or locating fixture. Tool-less assembly becomes increasingly difficult as the size and number of parts increases, and variations in locating points caused by fluctuations in temperature can be a problem. Tool-less assembly can reduce nonrecurring costs for low-rate production. The assembly process can be created to meet the form, fit, and function requirements for each product.

Embedded Sensors

Embedded sensors are sensors placed in a product to monitor performance and to provide feedback on adjustments. For example, embedded sensors in helicopter blades can determine the strain and lift in the blade, information that is then used to optimize performance. Sensors can also warn of conditions that require maintenance or repair. Problems with fabricating embedded sensors will have to be overcome, and sensors must become more robust and able to withstand harsher environments. Data transmission from sensors must also be improved. With advances in nanotechnology, microdevices and micromachines could be manufactured that would further the development of embedded sensors.

Flip Chips

Flip-chip technology (also called chip-on-board or direct-chip attachment) has the potential for electronic miniaturization, with up to 50 percent reductions in board area and 80 percent reductions in component weight over packaged devices, and increased reliability as a result of the elimination of one level of interconnection. Commercial applications include single-chip packaging, multichip packaging, and direct attachment to printed wiring boards. For use in advanced military applications, the thermal shock resistance, thermal cycling fatigue strength, temperature and humidity bias, and resistance to mechanical shock and vibration of flip-chip technology will have to be established.

Nanotechnology and Biotechnology

Nanotechnology, which involves the precise control of materials architecture at the molecular or atomic level, has great potential for the development of manufacturing processes that can vary material composition throughout a structure. Nanofabrication methods include nanomachining (in the 0.1 to 100 nm range) and molecular manufacturing (NRC, 1998). This technology could be used to manufacture microdevices and to produce complex shapes.

Biotechnology can precisely control molecular synthesis and assembly processes to produce a wide range of components from a limited number of constituent materials. Bioprocesses with potential manufacturing applications include: methods of coupling synthesis and self-assembly processes to produce oriented and functionally-graded structures; biological surfactant-based self-assembly processes that are effective in the 1 nm to 1,000 nm range; and biosynthetic pathways to genetically engineered protein polymers (NRC, 1998).

Environmentally Compatible Manufacturing Technologies

Cleaning Systems

Before a protective coating can be applied to a product, the surface of the product must be thoroughly cleaned. In the fabrication of electronics, several different surface layers must be cleaned to remove organic compounds that can affect adhesion. Contaminants, such as scale, must also be removed to reduce the risk of corrosion. In vapor degreasing, the most widely used method of cleaning, chlorinated fluorocarbons are heated and the vapor allowed to condense on the part being cleaned. This system is now being replaced, however, by systems that use cleaning solutions that are more environmentally compatible, but also more labor intensive. In airless degreasing systems, for example, parts are placed in a vacuum chamber and cleaned using freon vapor, which is then condensed and collected with limited exposure to the atmosphere. Ferrous materials and nonporous surfaces, such as castings, pose difficult challenges for these new cleaning systems. Cutting fluids used in fabrication processes are also being reconfigured so they will be easier to remove.

Coatings

Coatings are protective layers applied over parent materials to hinder corrosion or to protect them from exposure to high temperatures or other forms of energy. Coating processes include painting, chemical processing (e.g., anodizing), and the use of appliqués or stick-on coatings. Manufacturers have been shifting from the use of solvent-based paints to the use of water-based paints to decrease the environmental problems associated with application. In addition, controls are being put in place to trap the solids and volatile organic compounds generated during application of solvent-based paints. The use of paints containing chrome may soon be eliminated.

Problems associated with environmentally compatible coatings include the fact that water-based paints must be applied in lower humidity environments. As a result, water-based paints can only be applied to naval aircraft on an aircraft carrier about 10 days per year. From an environmental perspective, appliqués are considered to be a move in the right direction, although concerns about their durability remain. When process application and durability issues have been addressed, appliqués are expected to be used on many products, although cost competitiveness may still be a problem.

Materials Selection, Storage, and Disposal

Product materials must meet customer requirements, such as titanium, which is being used in military fighters because it has a greater strength-to-weight ratio than aluminum. Composite materials and ceramics are also increasingly being used to meet customer specifications. However, the process of obtaining the raw materials may have adverse environmental impacts. In addition, materials currently being used may be increasingly restricted in the future by environmental regulations, and alternative materials may have to be developed.

Although substantial efforts are being made to develop and use alternative materials, some manufacturing processes will continue to require hazardous materials. The most significant environmental concern for defense manufacturing is the storage and disposal of these hazardous materials. Storage and disposal sites are now licensed, and regulations are likely to become increasingly restrictive on storage and disposal facilities, which may be required to maintain special storage areas, provide safety training, and develop emergency management plans. These requirements will increase liability and insurance costs and encourage a just-in-time acquisition policy for hazardous materials.

Advanced Approaches to Business Organization

In the past 15 years, many business organizations have been reassessing their strengths and weaknesses and identifying "core competencies." As a result, fundamental and lasting changes are being made in the nature of business relationships.

Interorganizational Practices

Teaming among Organizations. Teaming is an effective organizational approach to the collective pursuit of a shared objective because teams combine the contributions of many individuals to accomplish a single objective. In the defense shipbuilding industry, for example, dramatically reduced production rates for submarines prompted the Electric Boat Corporation to propose a teaming arrangement with its competitor, Newport News Shipbuilding, to reduce the overall costs of new attack submarines. When independent companies compete for a share of the market, the free flow of information between companies is discouraged. Teaming requires neutral or common processes and business objects that encourage the exchange of information between organizations. Like virtual enterprises, teams are created and dissolved rapidly in response to business opportunities. The security of information systems, intellectual property, competition-sensitive business processes and practices, and risk sharing must be resolved for teaming to be successful.

Virtual Enterprises. Virtual enterprises are ad hoc organizations of individual, autonomous enterprises that have joined together for a specific purpose. In response to rapid changes in the business environment, organizations must reshape themselves quickly to exploit business opportunities. The enterprises involved in a virtual enterprise are typically dispersed geographically and have their own organizational infrastructures. High-speed communications have enabled the development of virtual enterprises worldwide.

Significant barriers remain, however, to the optimal functioning of virtual enterprises, including impediments to the rapid and full exchange of information such as a lack of robust information infrastructure and data exchange standards. Differences in organizational cultures can also create conflicts. The general understanding of organizational culture (i.e., what it is, its importance as an aid or impediment to the formation of an effective virtual enterprise) and methods of meshing or modifying cultures must be improved.

Long-Term Supplier Relationships. Some businesses attempt to encourage competition by utilizing a large supplier base and pitting one competitor against another to achieve the best terms. Others try to maintain long-term relationships based on common objectives with a few suppliers. As these relationships mature, suppliers are valued for more than cost and delivery; other valued attributes include quality, willingness to participate in the development process, and willingness to adopt the goals of the project. These suppliers often become adept at providing customized service to meet their customers' needs. Long-term supplier relationships can work well for a majority of supply items, with the exception of commodity items.

Intra-organizational Practices

High-Performance Organizations. High-performance organizations are characterized by fewer management layers, no vertical responsibility structures (silos), and team-based management systems (Peters, 1989). Harshman and Phillips (1993) describe traditional organizations as follows:

> A traditional organization functions as a top-down, authoritarian, control-based hierarchy. It is typically characterized by tightly held power; separate, often competitive functional departments; inadequate communication; control systems with quantitative, short-term, crisis-oriented time frames; and out-moded "carrot-and-stick" motivation systems.

High-performance organizations differ from traditional organizations in their guiding principles, structure, work roles, operating procedures, communication flow, and motivational methods. Although these characteristics are often referred to as the "soft side" of business theory, the potential for improvement through organizational change is significant.

A simple technique for moving toward higher performance is co-location of individuals engaged in a common pursuit. As information and communication technologies improve, co-location may be virtual, rather than actual. In high-performance organizations, information is credible and is freely shared. In addition, strict accuracy and configuration control are maintained.

Cross-functional Teams. Cross-functional teams bring together expertise from several areas to work toward a common objective. Organizations may use cross-functional teams as an alternative to serial function-to-function teams. This approach offers concurrency of input and is at the heart of the integrated product and process development approach. The practice of creating teams is still evolving, and many organizations have maintained functional groups to preserve process-specific knowledge and proficiency, although work products are generally developed by cross-functional teams.

Lean, Adaptive, and Agile Enterprises. Lean enterprises are characterized by only value-creating activities as defined by the customer and by processes that flow smoothly to meet the needs of the customer. In lean enterprises, all value-creating activities (from concept to product launch, from order to delivery, and from raw materials to the delivery of the product) are closely linked (Womack and Jones, 1996). Lean principles are increasingly integrated on the shop floor, but are being integrated more slowly in other areas of enterprises.

Adaptive enterprises are characterized by the delivery of high-quality products when the customer wants them with minimal resources and lead times. These enterprises are structured to apply resources (e.g., people, capital, suppliers) flexibly to meet customers' changing requirements. At the same time, they apply lean business principles to reduce cycle times, improve quality, and lower costs.

Agile enterprises provide customized products by working adaptively, responding to marketplace opportunities, exploiting technology, and entering into strategic alliances. Agile enterprises, which are knowledge driven, pull together the resources necessary to produce profitable products and services regardless of where they will be distributed (Goldman, 1998).

Knowledge-Based and Learning Enterprises. Knowledge-based enterprises are characterized by their ability to capture, store, communicate, and leverage employee knowledge by integrating individual and process-oriented information and knowledge across the enterprise. Learning organizations foster the creation of knowledge, continuous learning, and the implementation of best practices.

Information and Communications Technologies

Electronic Commerce

Electronic commerce, which can include electronic advertising of products and services, solicitations of proposals, contracting of services, and execution of services (in some markets), promises rapid, accurate business transactions at minimum cost. The spread of electronic commerce will depend on standardized infrastructure and business practices and will require improved capabilities for exchanging product data across diverse platforms.

Virtual Co-location of People

Information can be integrated via the co-location of the individuals necessary to accomplish a task. Virtual co-location uses technology to bring people into close contact with each other via conference calls, interactive Internet sessions, and video conferencing. Product development can be facilitated by virtual co-location, as individuals at remote sites can share in the development and realization of products. Visualization via three-dimensional modeling and rapid prototyping can also be used, and product data can be transferred to manufacturing sites to accelerate product delivery. Technologies for virtual co-location include intelligent agents to notify individuals when they are needed and mobile video and voice communications. Even though virtual co-location will allow geographically dispersed individuals to view and discuss the same information, it will be no substitute for physical presence for establishing and maintaining relationships between members of a high-performance team.

Data Interchange Standards

Data interchange standards enable data sharing across the platforms, CAD systems, and operating systems used by different organizations. Industry-consensus information models are under development for the shipbuilding, automotive, aerospace, and electronics industries. Currently, there are limited standards-based interchanges using ASCII text files. Software components to implement product data standards and libraries of product information will have to be developed.

Internet, Intranets, and Browser Technology

Web servers and hypertext markup language are used for publishing, the common gateway interface is increasingly being used to connect databases to intranets, and lightweight data access protocol is now shared among major software applications.

Intelligent Agents

Intelligent agents, built using modular artificial intelligence rule-based components, can perform simple tasks today. For example, desktop applications include "wizards" for performing routine tasks and e-mail applications include "personal assistant agents" for notifications and e-mail filtering. Mobile agents that can move from platform to platform in the performance of simple operations and learning agents that can adapt to the personal habits of users are being developed. Communities of these intelligent agents will soon be able to work together to perform more complex tasks.

Seamless Data Environment

In the seamless data environment, information sources will be transparent to the user, who will only need to know what information is required, not where to find it. Information has been stored in databases based on organizational structures, business processes, or other rationales, but access to information has been predicated on knowing where the information is stored and how it can be retrieved. Internet protocols connect systems within and across organizations and browsers gather information from sources unknown to the user. In a seamless data environment, information will be drawn from numerous sources within enterprises and from outside and will be presented to the user in a single presentation.

Telecommunications

Developments in telecommunications are moving toward an integrated public system that could make all forms of data and information easily and uniformly accessible throughout the world. The growth of the Internet and the World Wide Web has challenged communications carriers to increase network capabilities, with data traffic doubling every 14 months. Wireless satellite-based communication systems, land-based microwave systems, and other systems have been developed in response to the demand for flexible, mobile, worldwide communication systems. One example is the Iridium system, which combines 66 low-earth-orbit satellites with land-based wireless systems, as well as the capability of routing calls through land-based telephone systems if customers are within a local cellular coverage area.

Optical networking has the potential to meet demands for increased bandwidth and data transmission rates by carrying digitized data, voice, and video on multiple wavelengths of light. Optical fibers are rapidly replacing copper wire as a faster and more secure way of transmitting communication signals. Optical networking keeps communications traffic in an optical format. Data rates would increase significantly if communication messages were converted into an optical format and decoded into electronic format only on receipt by telephone, fax, or computer.

Distance Learning

Learning involves the transfer of knowledge and the ability to use that knowledge. Knowledge transfer can occur through a variety of media, including written materials, electronic media, and personal interactions. The ability to use knowledge requires repetition and feedback, which can be facilitated by interaction with an expert or mentor. Interactive sessions between experts and students can be provided via the Internet. Virtual reality and simulation techniques may eventually also enhance the learning process. In addition, an expert could be brought to a location virtually where his or her expertise is needed, such as medical teams performing complex medical procedures.

LEVERAGING COMMERCIAL ADVANCES

Although advances in the commercial sector can be leveraged to meet many aspects of the defense manufacturing challenges identified in Chapter 1, some required capabilities will have to be developed specifically by the defense community.

Low-Cost Rapid Product Realization

Opportunities

Rapid (and flexible) product realization refers to the ability to undertake low-volume production at a reasonable cost, as well as the ability to build defense products on commercial lines, customize products, and reconfigure products. In the future, defense products will have to be developed and manufactured more rapidly and at lower cost. This goal can be achieved by reducing cycle times and nonrecurring costs. Drastic reductions in cycle time and nonrecurring costs can be expected as a result of the following manufacturing advances:

- New approaches to manufacturing accounting. Using activity-based accounting, cost drivers can be more easily assessed. Using CAIV, performance and design criteria can be traded off to meet cost objectives. These advances have the potential to streamline and reduce costs in product realization processes.
- New approaches to product design. IPPD can reduce cycle times by reducing the need for redesigns late in the product realization process. In addition, standardizing parts and reducing the number of parts can reduce the cost of components and assemblies, as well as the need for new components. Three-dimensional digital product models can also reduce cycle time and late redesigns by predicting problems before physical resources have been committed. Simulation and modeling can also reduce cycle

times by revealing problems in processing before physical resources have been committed. With rapid prototyping, prototypes can be produced quickly from three-dimensional models.
- New approaches to manufacturing processes. GNC can be used to automatically generate the manufacturing plan or process concurrently with the three-dimensional data set so as to reduce flow time on the factory floor. Advances in soft tooling, flexible tooling, and tool-less assembly will enable low rate production and the production of different products with minimal cost and reconfiguration time. The assembly process can be created to the form, fit, and function requirements of the product.
- New interorganizational practices. As industries shift to teaming among organizations, virtual enterprises, and long-term supplier relationships, they will have access to a large base of potential manufacturers and will be able to develop, design, and produce products at the facility best suited to the task. Both costs and cycle times for product realization will be reduced. The merging of commercial and defense production lines would facilitate the production of weapons systems on largely commercial production lines.
- New intra-organizational practices. As organizations shift to high-performance, lean, adaptive, and agile enterprises and knowledge-based and learning enterprises, functionally integrated teams will drastically reduce production cycle times and costs, as well as overall product realization times. As enterprises become more agile and adaptive, they will be able to reconfigure rapidly to meet the requirements of new products and, consequently, reduce cycle times and costs. In addition, reductions in cycle time and lot size available from adaptive organizations will provide significant tools for low-volume production.

Gaps

The only gaps to be filled are in adapting these advances to the manufacture of defense-unique products. Defense organizations will have to undertake development initiatives for the production of composites, low-volume production, surge production capacity and capability, remanufacturing of parts and assemblies, customization of weapons systems, and the rapid reconfiguration of production lines to handle multiple defense products. Joint service development of weapons systems and technology exchange among programs and services would be helpful for decreasing cycle times and costs.

Expanded Design Capabilities

Opportunities

The design capabilities needed by defense manufacturing are: the design of products for multiservice use; designs that incorporate product life-cycle information; designs with extended-life in mind; designs for the maintainability of weapons systems; designs for technology insertion; open-architecture designs; designs for remanufacturing; designs for production by commercial processes; designs for the incorporation of COTS parts; designs for customization; and designs for reconfiguration. Advances in commercial manufacturing that could provide these capabilities are:

- New approaches to manufacturing design. Developments in simulation and modeling, including three-dimensional modeling, will enhance design capabilities by providing better representations of product performance as a function of design variables. Modeling and simulation of manufacturing processes and systems will facilitate the design of products for manufacturability. Significant advances are also anticipated in life-cycle design capabilities.
- New approaches to manufacturing accounting. The ability to design according to CAIV will also support this defense requirement.

Gaps

A number of design capabilities will not be achieved by advances in commercial manufacturing and will therefore require initiatives by the defense community. These include the integration of COTS into defense systems, multiservice functionality, extended-life weapons systems, improved maintainability, technology insertion, the customization, remanufacture, and reconfiguration of defense-unique products, and the use of commercial processes in defense manufacturing.

Environmentally Compatible Manufacturing

Defense manufacturing with low environmental impact, also called "green manufacturing," will be required to comply with increasing environmental constraints. In addition, depot and maintenance processes must have minimal environmental cost, and products must be designed using life-cycle analyses. The following advances have potential for meeting the needed capability of environmentally compatible manufacturing:

Opportunities

- New approaches to information technology. The exchange of product and process data will provide a vehicle for capturing and identifying environmental data related to specific products and processes.
- New approaches to manufacturing design. Life-cycle design technology will provide a tool for analyzing the environmental impacts of products at all stages in their life cycles. Through trade-off analyses, products can then be designed to minimize their environmental effects.
- New approaches to manufacturing processes. Developments in this area will be useful for the defense requirement of reducing environmental impact. Advances in coating and cleaning systems will be particularly advantageous for improving depot operations. Advances in material selection, storage, and disposal will also improve many defense products and processes.

Gaps

Some defense capabilities will not be addressed commercially, such as defense-unique coatings. The development of pollution abatement for defense-unique materials and chemicals must be ongoing.

Adaptation of Information Technologies

Opportunities

Defense manufacturing will need the capability to develop enabling technologies for specific applications, the capability to participate in the development of standards to ensure compatibility between defense and commercial systems, and the capability to develop product and process databases that incorporate design history, as well as worker rationale and know-how.

- New approaches to manufacturing processes. Improvements in the simulation of products and processes will generate information needed for defense databases and the rationale for particular designs or processes.
- New approaches to information technology. Standards for data interchanges and the exchange of product and process data will facilitate the use of information technologies for defense applications.
- New intra-organizational practices. In knowledge-based systems and organizations, rationale and know-how can be generated and captured for the design and manufacture of defense product and process databases.

Gaps

Government and defense contractors will have to invest in the development of simulation and modeling for defense products, in the development of cost models for all stages of product life cycles, in methods for ensuring data security,[1] and in the development of interoperable[2] commercial and defense systems.

Security of Product and Process Data

Opportunities

New approaches to information technology can support data security. Standards for data interchange, seamless product and process data flow, and the exchange of three-dimensional product models will make it easier to secure data.

Gaps

Although industry will develop systems for data security, they are not likely to meet the military's strict requirements, particularly for securing classified information. Therefore, security systems for defense product and process data will have to be developed, including explicit identification of suppliers.

Access to Production Sources

Opportunities

Defense manufacturing will need guidelines for commercial industry on critical components and subsystems, identification of suppliers, strategies for maintaining alternative suppliers, and adjustments in domestic source requirements to take advantage of foreign sources.

- New approaches to manufacturing processes. Flexible processes will increase the number of manufacturers for a given part. Flexibility will be aided by rapid prototyping, three-dimensional product models, high-speed machining, improved simulation and modeling, production process controls, adaptive machine controls, and tool-less assembly, as well as flexible tooling and soft tooling.

---

[1] "Data security" refers to the ability to protect military product specifications and other product information when these products are manufactured in locations outside of the well-defined U.S. defense industrial base.

[2] "Interoperability" refers to the ability to exchange files and link software and hardware systems between defense and commercial industries.

- New approaches to intra-organizational practices. The principles of agile enterprises will enable organizations to rapidly reconfigure production lines for new products, thus increasing the number of alternative manufacturers for a given product.
- New approaches to interorganizational practices. As industry's rapid teaming and communication with other enterprises improve, the capability to acquire alternate sources rapidly will also improve.

Gaps

Advances in product and process transportability and enterprise and process flexibility will increase the number of potential manufacturers and provide a hedge against a loss of suppliers through normal attrition or because of a military conflict. However, if a unique capability makes transportability difficult, DOD will have to have systems in place to identify alternate suppliers. DOD must also be able to identify critical components and specify sources of secure production, which will necessitate the identification of suppliers.

Use of Commercial Manufacturing Capacity

Opportunities

Defense manufacturing will need the capability to use commercial manufacturing capacity, including the use of and design for commercial processes, the incorporation of COTS parts and subsystems into defense products, the production of complete defense weapons systems on commercial lines, the reform of acquisition procedures to accommodate commercial practices, the monitoring of industry developments through technology road maps, the development of surge production capability, the avoidance of parts obsolescence, the qualification of commercial parts for military environments, and incentives for commercial industry to manufacture defense parts. Commercial manufacturing will have the ability to design some defense products for commercial production. The following advances will contribute to developing this capability:

- New approaches to manufacturing processes. All advances that increase flexibility for accommodating a wide range of products and configurations will facilitate the production of defense products on commercial lines, including rapid prototyping, improved simulation and modeling, tool-less assembly, and improved process control technologies. Increased process flexibility will also facilitate the development of surge production capacity and will enable more facilities to reconfigure production processes to accommodate the requirements of defense product. Flexible tooling and soft tooling can also enable the use of commercial manufacturing capacity.

- New approaches to information technology. Standards for data interchange will make it possible to exchange product and process data more efficiently and with greater freedom. Commercial standards will encourage defense agencies to submit their production requirements in commercially acceptable formats. The easy transfer of commercial process data to defense designers will help them accommodate commercial processes. Seamless access to product and process information will help.
- New approaches to manufacturing design. Comprehensive life-cycle design and design with CAIV will enable defense designers to trade off costs of commercial processes and design products for manufacture using commercial processes.
- New intra-organizational practices. As enterprises become more agile, they will be able to respond better to variations in customer requirements and more easily accommodate the special requirements of defense products. Methods of manufacturing small lot sizes will also enable the production of defense products. Defense manufacturers will probably move toward consolidating defense and commercial lines to take advantage of economies of scale.

Gaps

DOD should remove nontechnical barriers to the use of commercial facilities, such as outmoded accounting practices and acquisition regulations. Some defense capabilities not addressed include: the qualification of commercial parts to be used in defense systems that must withstand harsh environments; the frequent obsolescence of commercial parts; and the maintenance of overall system reliability with commercial parts. Continued support of lean and agile initiatives for defense contractors will be necessary until commercial organizations can meet defense requirements. Product and process requirements that impede the production of defense products by multiple facilities will have to be reduced. Technology transfer from commercial sources should be encouraged and incentives for commercial industry to manufacture defense products should be strengthened.

Sustainment of Weapons Systems

Opportunities

Weapons systems and other defense products will have to be longer lived than they have been in the past, as well as more fault-tolerant and more easily upgraded. They will require built-in diagnostic systems and more efficient techniques for routine maintenance. The capabilities needed by defense manufacturing include: life cycle analysis; extended life designs for weapons systems; designs for maintainability; designs for technology insertion; improved maintenance and depot processes; the development of remanufacturing processes; improved diagnostics; and product and process databases that include know-how. Relevant advances are listed below:

- New approaches to manufacturing processes. Advances that reduce assembly steps, the number of parts, and the number of interfaces (e.g., single-piece fabrication and high-speed machining) will produce weapons systems that are easier to sustain. Flexible assembly and forming processes will support the remanufacture of products. Embedded sensors will facilitate the development of improved diagnostic systems. Advances in manufacturing processes with low environmental impact will result in lower costs for depot operations and disposal of materials and products.
- New approaches in manufacturing design. Life-cycle design, the standardization of parts, and the reduction in the number of parts will simplify designs and facilitate easier maintenance and support. Cost trade-offs using CAIV accounting principles can facilitate the determination of life-cycle costs.
- New approaches to information technology. Seamless sources of product and process data and data exchange standards will support advances in product and process databases and remanufacturing capabilities, which will in turn support weapons system sustainment.

Gaps

Defense manufacturing capabilities that will not be met by the advances listed above include: extended-life designs for weapons systems; designs for maintainability; designs for technology insertion; more efficient maintenance and depot operations; the development of product and process databases; and improved diagnostics. DOD will have to support the development of methods for quantifying the ability of commercial parts to withstand harsh military environments. Defense manufacturing should be proactive in monitoring advances in commercial technology and planning for their incorporation. Recent industry road maps should be used as one source of information. DOD should also consider the management of the supply chain and establishing incentives for commercial manufacturers to produce defense parts. Although advances will be made in life-cycle analysis methods, DOD will have to develop the aspects of analysis and life-cycle models that are peculiar to weapons systems, including long-lived systems and systems that must operate in harsh environments.

SUMMARY

Pressures are increasing on defense manufacturing to make use of commercial manufacturing advances, products, and production capacity. In addition, the commercial sector can provide a number of opportunities. Currently, DOD is actively pursuing the use of commercial production lines for the manufacture of defense products. The ManTech program has established several pilot projects to establish feasibility, and an Air Force pilot project has demonstrated production, on a commercial automotive manufacturing line, of digital electronic modules compatible with the F-22 Raptor fighter and the RAH-66 Comanche helicopter (Heberling et al., 1998).

The use of COTS products for defense systems holds great promise but also raises concerns. First, a supplier could stop manufacturing a product if the commercial market for it becomes unprofitable. Commercial products tend to have much shorter lives than defense products and tend to be replaced with new technology much more often. In addition, it will be necessary to guard against situations in which manufacturers abandon a commercial product that has been designed into a defense system. To address this problem, design systems will have to be technology transparent and based on modular open architectures to permit new commercial components, technologies, and functions to be used to upgrade defense systems. The F-22 Raptor program is a case in point. Originally designed for existing components, millions of dollars were spent to redesign the system to accommodate new electronics components, even before the plane had entered production. Shortening the development and product realization cycle would also help avoid such problems.

Second, the design limits for commercial applications may be exceeded in military use. The military environment can be harsher than the commercial product environment (e.g., high acceleration forces, vibration, and corrosive conditions). DOD will have to qualify commercial parts that are not specifically designed to withstand these environments and, if necessary, modify them to meet military needs or develop system designs that compensate for the limitations of commercial parts.

The production of military parts on commercial production lines also raises concerns. Because system, subsystem, and component designs would have to be appropriate to modern commercial processes, defense manufacturers must keep abreast of advances in commercial processes, accommodate them in their designs, and pursue enabling technologies and practices that would facilitate the use of commercial production lines. Table 3-2 summarizes the defense manufacturing challenges that are supported by commercial advances.

TABLE 3-2 Defense Manufacturing Challenges Supported by Commercial Advances

Challenge	Supporting Commercial Advances	Elements
Low-cost rapid product realization	Industry collaboration High-performance organizations Adaptive enterprises Advanced manufacturing processing technology	Activity-based accounting Cost-as-an-independent-variable accounting Integrated product and process design Three-dimensional digital product models Simulation and modeling Tool-less assembly Teaming among organizations Virtual enterprises Long-term supplier relationships Lean, adaptive, and agile enterprises Knowledge-based and learning enterprises Simulation and modeling
Expanded design capabilities	Life-cycle perspectives Advanced manufacturing processing technology	Simulation and modeling Three-dimensional digital product models Life-cycle design Cost-as-an-independent-variable accounting
Environmentally compatible manufacturing	Shared information environments Life-cycle perspectives Environmentally compatible manufacturing	Seamless data environment Life-cycle design Coating systems Cleaning systems Material selection, storage and disposal
Adaptation of information technology	Shared information environments Adaptive enterprises Advanced manufacturing processing technology	Simulation and modeling Data interchange standards Seamless data environments Knowledge-based enterprises Simulation and modeling
Security of product and process data	Advanced manufacturing processing technology	Data exchange standards Seamless data environment Three-dimensional product models

continued

TABLE 3-2 continued

Challenge	Supporting Commercial Advances	Elements
Access to production sources	Industry collaboration Shared information environments Adaptive enterprises Advanced manufacturing processing technology	Rapid prototyping Three-dimensional product models High-speed machining Simulation and modeling Adaptive machine controls Tool-less assembly Agile enterprises Teaming among organizations
Use of commercial manufacturing capacity	Shared information environments Life-cycle perspectives Adaptive enterprises Advanced manufacturing processing technology	Rapid prototyping Simulation and modeling Tool-less assembly Data interchange standards Seamless data environment Life-cycle design Cost-as-an-independent-variable accounting Agile enterprises
Sustainment of weapons systems	Life-cycle perspectives Advanced manufacturing processing technology Shared information environments Environmentally compatible manufacturing	High-speed machining Embedded sensors Cleaning systems Coating systems Life-cycle design Cost-as-an-independent-variable accounting Seamless data environments Data interchange standards

4

New Priorities for Defense Manufacturing

Defense manufacturing will face major challenges between now and 2010. At the same time, defense manufacturing will have many opportunities to develop innovative manufacturing methods and technologies that promise higher efficiency, lower costs, and greater capabilities than ever before. The expected interdependence of commercial and defense manufacturing is especially promising. Meeting the challenges of defense manufacturing in 2010 will require a new focus on commercial markets, which will reshape the priorities and organizations of both defense manufacturers and federal defense agencies. Chapters 1–3 described the challenges facing defense manufacturing, the manufacturing capabilities required to meet defense needs in 2010, and the potential for meeting these needs by leveraging commercial advances. This chapter reviews these challenges and offers recommendations for meeting them.

SETTING PRIORITIES

Cost-effectiveness must have the highest priority in future defense manufacturing requirements because of the expected continued decline in the defense budget. The committee believes that the principal criterion for prioritizing manufacturing capabilities for development and investment should be potential cost savings (e.g., return on investment). In this report, the committee has applied this criterion by emphasizing capabilities that (1) will be broadly applicable to many weapons systems or many elements of life-cycle costs; (2) will benefit from substantial nondefense resources; (3) will address large expenditure budget items for DOD; (4) could lead to significant performance or productivity gains; (5) will address problems likely to become more important in the future; or (6) will not be developed as a result of commercial investment. The committee concludes that the following four categories of defense manufacturing capabilities offer the greatest potential returns on investment:

- efficient sustainment of weapons systems
- modeling and simulation-based design tools
- leveraging of commercial resources
- cross-cutting defense-unique production processes

The committee recommends that current DOD research and development efforts in defense manufacturing be augmented in these four high-priority categories. In the following sections, the committee recommends several areas within these categories for development.

Efficient Sustainment of Weapons Systems

Research and development priorities in the efficient sustainment of weapons systems should be focused on reducing sustainment costs by shortening product cycle times and developing low-cost processes for maintenance and repair, improving the reliability of new and existing weapons systems, and upgrading new and existing systems. In 1997, the DOD budget for operations and maintenance was approximately twice as large as the budget for procurement and represented approximately 36 percent of the total defense budget (OMB, 1998). Sustainment represents a significant fraction of the life-cycle costs associated with the operation of weapons systems. Because the proportion of aging weapons systems in the inventory will continue to grow, the problem will become more difficult by 2010 and will continue to consume a significant portion of defense resources. Many sustainment capabilities (e.g., improved diagnostics, open architecture, parts logistics, depot floor operations, and remanufacturing) are applicable to many weapons systems, so improvements would have broad applicability and large benefits. The potential for improving sustainment is significant because many modern manufacturing concepts (e.g., lean manufacturing) have not been widely applied to depot and maintenance operations. Commercial manufacturing is unlikely to provide the needed capabilities because few commercial industries have such long-lived product lines, and many of the issues related to the sustainment of weapons systems are defense-unique.

Recommendation. Current and future DOD manufacturing research and development aimed at improving sustainment capabilities for aging weapons systems should emphasize the following areas:

- *Application of advanced production processes and practices to*

maintenance, repair, and upgrade operations. Many recent improvements in the cycle time and productivity of manufacturing operations are applicable to sustainment operations. Areas of particular importance include shop floor process controls, new organizational structures (e.g., self-directed work teams), reductions in cycle times, management of inventory, and continuous flow manufacturing.
- *Technology insertion for new and existing systems.* New weapons systems should be designed with open architectures and should be "technology transparent" (i.e., upgradeable without major redesigns). Better ways to incorporate COTS products into aging systems without replacing major subsystems should also be investigated to reduce the cost of inserting new technologies. In addition, new system designs should be developed with the goal of increasing system reliability significantly over existing systems.
- *Self-diagnostics for mechanical and electronic systems.* Intelligent monitoring systems should be developed that can detect current problems and assess the probability of future failures. The predictive approach goes beyond the built-in test capability of current electronic systems. For mechanical systems, this capability could be achieved through sensors that monitor cumulative stress and structural reactions. Advances in the miniaturization of components (e.g., microelectromechanical systems) may be useful for this application.
- *New technologies for remanufacturing.* Methods of noncontact gauging for rapidly capturing mechanical part geometry should be explored. Programmable free-form processes for rapid remanufacturing should be developed.
- *Design methods that improve sustainment.* Methods and capabilities should be developed to incorporate total life-cycle, maintainability, high reliability, and technology insertion into new weapons system design processes.

Modeling and Simulation-Based Design Tools

A new, more powerful design environment is evolving, with the capacity to predict the performance and manufacturability of products early in the design process. This simulation-based environment will allow design trade-offs to be made at the conceptual stage, as well as at the detailed design stage, and will permit the early optimization of life-cycle costs. (Design changes at the conceptual stage will have the greatest effect on product costs.) Simulation-based design will enable the concurrent design of products, manufacturing processes, and maintenance procedures. Better use of information technology in the design process would enable designers to take into consideration commercial developments in modeling and simulation, database search engines, product data structures, and distributed design methods.

Recommendation. DOD should further encourage defense industry efforts to make the most of the simulation-based design environment and should focus on the following activities:

- promote the development of models of defense products, manufacturing processes, and life-cycle performance
- develop algorithms for design trade-offs that optimize life-cycle costs
- develop enhanced and easily usable parametric models that facilitate design trade-offs at the conceptual stage
- initiate the development of product databases that will permit simulation at various levels of resolution

Leveraging of Commercial Resources

The commercial manufacturing industry will continue to drive innovations in manufacturing technology simply because of the size of its investments compared to those of defense manufacturing. As the distinction between commercial and defense industries lessens, defense manufacturing can benefit from adopting the "best practices" of commercial industries. Commercial developments, to the extent that they lower the life-cycle costs of products, will tend to reduce the pressure on defense procurement and operations and maintenance budgets.

Recommendation. Advances in commercial manufacturing should continue to be monitored and adapted to defense applications as appropriate. Technology road maps created by commercial industry should be used to help defense manufacturing programs keep abreast of developments and forecasts.

The increasing use of COTS products can dramatically reduce the costs and development cycle times for defense products. Most DOD acquisition requirements for new weapons systems require COTS hardware or software whenever feasible. The committee believes that a strategy to incorporate COTS products into existing weapons systems should also be pursued. Even though inserting COTS products into existing systems is not as straightforward as using them in new systems designs, they could significantly reduce costs, especially in light of the growing numbers of aging systems.

Recommendation. The following development areas should be pursued to facilitate the widespread use of COTS products:

- new weapons systems designed for open architecture and technological transparency
- a central program and mechanisms to maintain awareness of, document, and plan for new COTS technologies that can be incorporated into current

and future weapons systems, as well as to disseminate this information to individual program offices
- improved methods of inserting COTS products in fielded weapons systems
- low-cost validation methods for determining the adequacy of COTS parts for military applications

Defense-Unique Production Processes

The committee identified a variety of production processes that are applicable to many defense products. These processes are generally defense-unique, although there is some overlap with commercial processes.

Recommendation. Defense manufacturing programs should continue to address the development and improvement of defense-unique and defense-critical processes. The following defense-unique and/or defense-critical processes have the broadest range of applications:

- processes that enable rate-transparent production (i.e., production where the per unit cost is independent of the production rate), including programmable free-form processes (no hard tooling); easily reconfigurable production lines (to permit production of different products on the same line), simulation models and tools for production systems (to optimize processes prior to the commitment of physical resources), and adaptive process controls (to increase first-time yield)
- processes for the low-cost fabrication of composite structures, including automated fiber placement for complex shapes, rapid autoclave processes or nonautoclave processes, and automated structural repair processes
- processes for the low-cost production and application of low observability coatings and structures, including automated coating processes for obtaining uniform and accurate coatings on complex shapes, coating thickness sensors that can operate in severe process environments, forming processes for complex shapes, processes to coat interfaces between parts, and designs to eliminate interfaces between parts
- defense-unique electronic technologies, including packaging for harsh environments, integrated systems-on-a-chip, flip chips, multichip packaging, and rugged, uninterruptable interconnections that can operate in severe vibration environments
- design, information, and manufacturing technologies that provide dimensional control in the production of large, complex parts

REORIENTING PROGRAMS

DOD's ManTech program, a joint program of the armed services and the

Defense Logistics Agency, focuses on the development of manufacturing technology for the affordable, low-risk development and production of weapons systems. The objective of the ManTech program is to link technological innovations and developments with production.

The six thrust areas of the ManTech program are: (1) metals processing and manufacturing, which focuses on developing affordable, robust manufacturing processes and capabilities for metals and specialty materials that are critical to defense applications; (2) composites processing and manufacturing, which promotes the production of composite structures that can compete with metal structures in both performance and cost; (3) electronics processing and manufacturing, which concentrates on manufacturing technology for electronic materials, devices, integrated circuits, subassemblies, and subsystems; (4) advanced industrial practices, which encourages the adoption of the world's best practices in design, development, production, and life-cycle support of defense products; (5) manufacturing and engineering systems, which concentrates on manufacturing systems technology; and (6) sustainment/readiness, which focuses on improving readiness and logistics support. According to the 1998 budget and five-year budget projections, the thrust areas related to production processes (metals processing and manufacturing, composites processing and manufacturing, and electronics processing and manufacturing) receive about 70 percent of the annual funding; advanced industrial practices receives 20 percent; and manufacturing and engineering systems and sustainment/readiness receive 5 percent each. The projects within these thrust areas are usually directed toward specific program applications rather than generic technology development because the weapons system program managers for acquisition and logistics are considered the primary customers for the ManTech program. Because of this program orientation, the emphasis has been on coordinating program advances and technology implementation across the spectrum of defense manufacturing.

The committee believes that the ManTech program is an ideal vehicle for developing many of the required defense manufacturing capabilities described in this report. However, the program needs new directions, including new thrust areas, to meet future demands.

Recommendation. The ManTech program should focus on the following roles to meet the needs of defense manufacturing in 2010. (Some of these roles require only a change in emphasis of existing roles, but some are new roles that should be incorporated into the program charter.)

- *Leader in affordability.* The ManTech program should be a primary means of achieving weapons systems affordability throughout the life cycle. To serve in this role, the ManTech program should broaden its focus to include the front end (conceptual design and development) and the back end (sustainment) of weapons system life cycles. In addition, the

ManTech program should take a more proactive role in executing projects with large financial impacts on system costs and a more aggressive approach to dispersing new technologies across services and weapons systems.

- *Focal point for cross-cutting defense technologies.* The ManTech program should focus on projects whose results are expected to be widely applicable and on minimizing the duplication of projects by individual program offices. Focusing on cross-cutting technologies could lead to substantial cost savings.
- *Technology middleman.* The ManTech program should promote the implementation and dissemination of new technologies. In this role, the ManTech program would provide advice and assistance about future technologies for defense program management offices and the industrial design teams responsive to them. One mechanism for technology dispersion would be the temporary transfer of ManTech personnel to program offices or defense contractors as members of design teams or integrated product and process development teams. The goal would be to ensure that new technologies are accepted in the development, production, and support of weapons systems.
- *Information broker and planner.* The ManTech program should expand its role in providing information on new technologies by monitoring commercial technology developments so plans can be made for proactively incorporating them into defense systems. The ManTech program should distribute these plans, along with information on new technologies, to the defense community. Industry road maps are one important source of this information. In addition, ManTech development program time horizons should be extended from the current 5 years to about 10 years so that technologies with the potential for significantly affecting weapons systems can be implemented. At present, manufacturing technology projects are often planned so that payback will occur during the development cycle of a program (i.e., within five years). Many significant manufacturing technologies, such as simulation-based design and adaptive manufacturing systems, have the potential to make significant advances in the next 10 to 15 years. Because not all projects need these longer time horizons, however, the ManTech program should maintain a balance of short-term and long-term projects.
- *Expert in weapons systems technologies.* The ManTech program should develop a greater understanding of technologies that are important to major weapons systems to facilitate the implementation of new technologies. It might be useful to assign personnel with extensive experience from systems program offices and the defense industry to the ManTech program.

Recommendation. The ManTech program should consider revising its division of effort if it is to implement the new roles and development initiatives that the committee has recommended. The following changes are recommended:

- *Production processes.* There are many more opportunities to improve defense-unique production processes than can be accomplished within budget constraints. The committee recommends that production processes continue to be a major thrust area, but the emphasis should be shifted toward cross-cutting technologies. Two examples of production process areas that lend themselves to cross-cutting technology development are composites and electronics. In addition, the ManTech program should expand its efforts to find multiple sponsors for projects to encourage the widespread application of new production processes.
- *Advanced industrial practices.* Adopting industrial best practices will continue to be important for defense manufacturers. However, as defense contractors integrate commercial and defense production and more defense subsystems are manufactured on commercial lines, defense manufacturers will naturally adopt best industrial practices on their own initiative. The ManTech program should expand this area beyond best practices to include technologies for enhancing cost-effectiveness.
- *Manufacturing and engineering systems.* The ManTech program should establish an initiative for the development of simulation-based design tools in this thrust area. Even though the ManTech program has combined the areas of advanced industrial practices and manufacturing and engineering systems, the emphasis recommended for design should be maintained. The level of emphasis in this area should be at least as great as the development of production processes because of the leverage that simulation-based design can have on weapons system costs.
- *Sustainment of weapons systems.* The enormous opportunities for cost savings by new approaches to sustainment suggests that this thrust area should be greatly expanded, with an emphasis comparable to that of the thrust areas related to production processes. Primary project areas should include use of COTS products, self-diagnostics, and shop floor control processes.
- *Leveraging of commercial resources.* The ManTech program should establish a thrust area specifically directed at leveraging commercial resources. The keystone in this area is the use of COTS products, including designing for implementation, designing for technology transparency, and the validation or modification of commercial parts.

SUMMARY

The committee has recommended several major development initiatives intended to improve the cost-effectiveness of defense manufacturing in a high-priority class of manufacturing capabilities. The committee has also recommended new roles for the ManTech program that would make it more proactive and effective in dispersing technology to users. Finally, the committee has recommended changes of emphasis and direction in the six ManTech thrust areas and the addition of a thrust area for leveraging commercial resources. The committee believes that critical mass can be maintained in the important ManTech initiatives currently under way while the reorientation proceeds. Investments in the ManTech program already provide a return through cost savings and cost avoidance. With the recommended emphasis on projects and technologies with broad applicability, future returns on investments should be even larger.

References

Brooks, H., and B.R. Guile. 1987. Overview. Pp. 1–15 in Technology and Global Industry, H. Brooks and B.R. Guile (eds). Washington, D.C.: National Academy Press.

Congressional Budget Office. 1997. Reducing the Deficit: Spending and Revenue Options: A Report to the Senate and House Committees on the Budget. Washington, D.C.: Government Printing Office.

Defense Science Board. 1997. Report of the Task Force on Vertical Integration and Supplier Decisions. Washington, D.C.: Office of the Secretary of Defense.

DTAP (Defense Technology Area Plan). 1997. "1997 Defense Technology Area Plan." [Online]. Available: HtmlResAnchor http://www.dtic.mil/dstp/97_docs/dtap/dtaps.htm/ [1998, November 23].

Dertouzos, M.L, R.K. Lester, and R.M. Solow. 1989. Made in America: Regaining the Productive Edge. Cambridge, Mass.: Massachusetts Institute of Technology Press.

Dowdy, J.J. 1997. Winners and losers in the arms industry downturn. Foreign Policy 107: 88.

Friedman, A., J. Glimm, and J. Lavery. 1992. The Mathematical and Computational Sciences in Emerging Manufacturing Technologies and Management Practices. Philadelphia, Pa.: Society for Industrial and Applied Mathematics.

Goldman, S.L. 1998. Agile Manufacturing in 2010. Briefing presented to the Committee on Defense Manufacturing in 2010 and Beyond, at the National Research Council, Washington, D.C., February 2, 1998.

Harshman, C., and S. Phillips. 1993. Teaming Up: Achieving Organizational Transformation. San Diego: Pfeiffer and Company.

Heberling, M., J.R. McDonald, R.M. Nanzer, E. Rebentisch, and K. Sterling. 1998. Using commercial suppliers: barriers and opportunities. Program Manager 27(4): 48–55.

ManTech (Manufacturing Technology Program). 1998. "Department of Defense Manufacturing Technology Program." [Online] Available: HtmlResAnchor http://mantech.iitri.com/ [1998, November 24]

NGM (Next Generation Manufacturing). 1997. Next-Generation Manufacturing: A Framework for Action. Bethlehem, Pa.: Agility Forum.

NRC (National Research Council). 1995. Information Technology for Manufacturing: A Research Agenda. Washington, D.C.: National Academy Press.

NRC. 1997. Aging of U.S. Air Force Aircraft. Washington, D.C.: National Academy Press.

NRC. 1998. Visionary Manufacturing Challenges for 2020. Washington, D.C.: National Academy Press.

NSTC (National Science and Technology Council). 1997. Manufacturing Infrastructure: Enabling the Nation's Manufacturing Capacity. Report of the Subcommittee on Manufacturing Infrastructure. Washington, D.C.: The White House.

OMB (Office of Management and Budget). 1998. Budget of the United States Government: Historical Tables. Washington, D.C.: U.S. Government Printing Office.

ONI (Office of Naval Intelligence). 1995. Worldwide Submarine Proliferation in the Coming Decade. Washington, D.C.: Office of Naval Intelligence.

Peters, T., 1989. *Excellence in the Public Sector*. Boston, Mass.: Northern Light Productions. Videotape.

Rothschild, M., M.W. Horn, C.L. Keast, R.R. Kunz, V. Liberman, S.C. Palmateer, S.P. Doran, A.R. Forte, R.B. Goodman, J.H.C. Sedlacek, R.S. Uttaro, D. Corliss, and A. Grenville. 1997. Photolithography at 193nm. Lincoln Laboratory Journal 10(1): 19–34.

Stevens, D., B. Davis, W.L. Stanley, D.M. Norton, R. Starr, D.P. Raymer, J. Gibson, J.P. Hagen, and G. Liberson. 1997. The Next Generation Attack Fighter: Affordability and Mission Needs. Santa Monica, Calif.: Rand Corporation.

West, L.A. 1998. Product Realization Environment. Briefing presented to the Committee on Defense Manufacturing in 2010 and Beyond, at the National Research Council, Washington, D.C., February 2, 1998.

Womack, J.P., and D.T. Jones. 1996. Lean Thinking: Banish Waste and Create Wealth in Your Corporation. New York: Simon and Schuster.

Appendices

APPENDIX A

Historical Perspective on the U.S. Defense Industrial Base

Historical events have always influenced the United States' perception of its defense needs, which, in turn, have influenced the nation's commitment to maintaining defense supplies and manufacturing capacity. Participation in the French and Indian War (1750–1770) taught the colonists the importance of maintaining arms and the capability of using them when necessary. Prior to declaring independence in 1776, the colonists had gathered military supplies at a number of locations, some of which were used at Lexington and Concord after the "shot heard round the world" was fired. Immediately after achieving independence, while operating under the Articles of Confederation, the states established and equipped militia to ensure the integrity of their boundaries. Official purveyors of powder and guns were established, as well as locations for storing weapons and supplies. States with long coastlines established their own seagoing defense forces, procuring some vessels from shipyards that had been operating along the Atlantic coast before independence. These munitions storage locations and maritime construction yards were the first elements of the U.S. "defense industrial base."

PRE-WORLD WAR I

Before World War I, U.S. defense policy was still based on George Washington's philosophy of avoiding foreign entanglements. Except for a few occasions when it was necessary to send forces to protect U.S. interests abroad, the military's mission was perceived to be protecting U.S. borders from direct attack. Arsenals had been established to develop ground weapons appropriate to that task, and naval shipyards were responsible for providing vessels for U.S. naval forces to protect the coasts. Arsenals and shipyards devised, established, and operated defense manufacturing processes within their facilities. From 1885 to 1914, the United States enjoyed unprecedented industrial growth and economic prosperity, and new technology was used for the manufacture of large quantities of consumer and industrial goods rather than military weapons.

The best known developments in manufacturing technology that became important during World War I were based on the work of individuals, such as Samuel Morse, Alexander Graham Bell, Thomas Edison, Guglielmo Marconi, and Henry Ford. The use of wires to transmit messages over long distances, first in code and then by voice, put the United States in the vanguard of technical nations and expanded military communications capability beyond semaphore and messengers. Edison's development of the electric light bulb and the "talking machine" made him an internationally-known inventor. Marconi's wireless transmission devices, in turn, transcended wire-based technology and transferred information by "radio," a capability of tremendous military significance in the naval activities of World War I. Manufacturing methodology and practices developed alongside these advances. Ford's production line, which enabled the continuous large volume production of complex products, had the most profound effect. Production lines set the stage for the manufacture of complex defense systems on a grand scale.

WORLD WAR I

Initially, the United States acted as a supplier of military capability to other nations during World War I. However, support for Great Britain and France led first to U.S. assistance in North Atlantic convoy patrols (protecting maritime interests), then to supplying military equipment and other commodities to allied nations, and, finally, to U.S. entry into the conflict with the dispatch of the American Expeditionary Force to Europe. Mass production, already installed in many U.S. facilities, made it possible for the United States to augment allied industrial capacity by providing arms, ammunition, and some military vehicles. The United States had no capability, however, to produce or provide combat aircraft. The United States used foreign-manufactured aircraft in combat, with the exception of the Curtiss JN-2, or "Jenny," which was used extensively to train pilots who later engaged in air combat as members of other national forces or with the American Expeditionary Force in France.

1919 TO 1938

From 1919 to 1938, the U.S. military role was mainly to protect U.S. interests throughout the world. The disarmament begun in 1919 fostered hopes of peace throughout the world, and the League of Nations was widely regarded as a mechanism for overseeing an extended period of peace and prosperity. In 1929, this vision was undercut by the onset of the worldwide depression. First the United States, then Europe, was badly shaken by such severe economic turmoil that many questioned the very foundations of democratic government. In Italy, Mussolini and his Fascists assumed power and then embarked on aggressive military campaigns on the African continent. In Germany, Hitler's National Socialists began to consolidate their power. These nations rapidly expanded their industrial capabilities for producing armaments under the guise of producing commercial items (e.g., typewriters, baby carriages, glider aircraft). Emphasis was placed on the development of aircraft, and German factories were built based on new manufacturing concepts. The number of naval vessels constructed sharply increased, breaching the treaties limiting the size of naval forces. In Japan, mass production was introduced, and modern munitions and arms production facilities were established. During this period, munitions and other specialized military products were designed and produced within the U.S. armed services. Support for these weapons systems was provided by depots, arsenals, and shipyards in conjunction with the service supply and maintenance systems. These facilities can be considered a defense industrial base, although many weapons and weapons systems (e.g., aircraft for the Army Air Corps) were designed and produced in privately-owned plants. Although the Navy owned and operated two aircraft production facilities, many naval aircraft were also produced outside the defense industrial base. State-of-the-art manufacturing methodologies were used by both commercial and military facilities.

WORLD WAR II

In 1938, Congress chartered the Defense Plant Corporation, which, in anticipation of hostilities, was assigned the task of expanding production capabilities for military equipment. Its charter permitted both the building and equipping of new facilities and the expansion of existing facilities. It also had the authority to enlist the help of industrial organizations in establishing and operating facilities in the public interest. U.S. involvement in World War II began in 1940 when, under the Lend-Lease Program, the Roosevelt administration provided 40 World War I destroyers to Great Britain for use on North Atlantic convoy routes.

From 1939 to the end of World War II, the Defense Plant Corporation built many government-owned, contractor-operated (GOCO) facilities, the preponderance of which are either still operating or are on inactive standby. At the same time, arsenals and navy yards were expanded and worked two or three shifts a day producing weapons. With the help of government agencies, many U.S. industrial sectors converted to military production, incorporating new manufacturing methodologies that enabled the massive production of war equipment. Automobile and truck production lines were converted to military production, existing commercial shipyards were expanded, and new ones were built. In some commercially owned and operated shipyards, a new class of vessel—the liberty ship—was produced in massive quantities using newly developed manufacturing concepts. By the end of the war, Kaiser Industries was able to build a liberty ship in one day at its shipyard in Richmond, California.

Aircraft production was improved through new manufacturing technologies. Taking advantage of mass production techniques for automobiles, Ford built a plant at River Rouge, Michigan, to build B-24 bombers designed by the Consolidated Aircraft Corporation. Production throughout the aircraft industry soared with bomber production ultimately reaching the level of one bomber per hour. By early 1945, the combined capability of converted commercial plants and GOCO plants to produce military goods was truly awe-inspiring, and mass production manufacturing technologies had been significantly advanced.

1945 TO 1950

With the defeat of the Axis Powers, the Alliance, with the exception of the Soviet Union, began dismantling its military production capability and redirecting it to the production of consumer and commercial goods. A strong economy was considered the best defense, and most commercial facilities that had been converted to the production of military equipment reverted to the production of consumer and commercial products, while many GOCO plants were either downsized or closed down.

Perhaps the most far-reaching defense-related change in the postwar period was the reorganization of the military establishment and its associated civilian agencies. Wartime experience had shown that an integrated military organization was necessary to prepare for and fight a modern war and to coordinate land, sea, and air forces using common military equipment. On July 26, 1947, President Truman signed into law the National Security Act and Executive Order 9877. The Executive Order implemented the act and set forth the functions of all elements in the newly created U.S. Department of Defense (DOD). The new defense establishment faced not only the turbulence of a major reorganization and realignment of responsibilities, but also had to contend with a continuing reduction of forces as thousands of reservists were involuntarily discharged from the services.

In 1948, President Truman directed Secretary of Defense Louis Johnson to perform a complete review of defense needs, which resulted in further cuts in the already demobilized military base. Many units were dis-established and their equipment sent to storage areas. Aircraft were sent to Davis Monthan Air Force Base in Arizona, and ships were anchored in Mobile and Chesapeake Bays, the Hudson and James Rivers, and the Bremerton and Philadelphia navy yards. Some land combat equipment was dispersed to storage areas, but much of it was sold for scrap. The need for new equipment was minimal, and the significantly downsized defense industrial base was easily able to fill the demand. This retrenchment program was euphemistically known as the "Johnson Axe."

Paradoxically, during this period of demobilization, new defense technologies and manufacturing methods, especially in the area of aeronautics, were introduced in some of the armed services. The Air Force and Navy progressed from propeller to jet aircraft and developed several new military aircraft systems. Manufacturing methodologies and tooling in aircraft plants kept pace with the modernization trend. The Army, Marines, and Navy, however, continued to use World War II-era weapons and weapons systems.

KOREAN WAR

By 1950, the defense industrial base, at its weakest point since the mid-1930s, was generally not prepared to meet the greatly accelerated demands brought on by the Korean War. In terms of aircraft-related research, development, and manufacturing, the defense industry had been reasonably well modernized or was modernizing, but this was not true in other areas of defense materiel production. Most U.S. military forces had to make do with World War II equipment for the first several months of the war.

Congress quickly responded to the crisis and passed legislation encouraging and facilitating industrial expansion to meet projected military needs. The most important piece of legislation, signed into law by President Truman, was the Defense Production Act of 1950, which has been extended several times since and is still in effect today. The act (1) defined the defense industrial base; (2) established a priority system for obtaining necessary military hardware and software during emergencies; and, perhaps most importantly, (3) provided for seed money to establish quantity production of new defense materiel and to increase production capacity for specific equipment through the improvement of production methodologies and facilities.

In response to the Soviet development and introduction of the MIG series of jet fighters, the propeller aircraft of World War II were replaced by new jet aircraft. In addition, engine manufacturing methodologies and processes were updated to produce the Century Series of jet fighters.

1953 TO 1972

After the Korean conflict, the Cold War between the United States and its allies and the Soviet and Chinese blocs intensified, with the shift in political perspective greatly affecting the structure and mission of military forces. The emphasis shifted from preparing for tactical warfare to preparing for strategic warfare. Under the Single Integrated Operations Plan, planning and targeting became joint functions with all strategic forces placed under the operational control of a single commander in chief in the event of war. These strategic forces consisted of land-based bombers, land-based intercontinental ballistic missiles, and submarine-borne ballistic missiles, and were known as the "Triad." The development of nuclear weapons and their delivery vehicles was given high priority, with defense manufacturing focused on providing the requisite equipment.

In 1958, after the launch of Sputnik, the development and production of space systems burgeoned. Most U.S. space systems were designed to contribute to the country's overall nuclear deterrence by providing intelligence and early warnings of attacks and performing other command, control, and communications functions. Because spacecraft could not be produced using manufacturing systems designed for nonspace equipment, many new manufacturing processes, methods, and tools were developed. Extended operations in space created a need for metals that exceeded previous levels of purity. As the U.S. space program was accelerated toward the promised moon landing, pressure increased to investigate the effects of prolonged orbital and extra-terrestrial activities on equipment and human beings.

The conflicts in Vietnam, Laos, and Cambodia reversed these priorities for a time. Industry was called on to shift gears once again and produce more conventional weapons and weapons systems, while continuing to meet equipment requirements for nuclear deterrence. As the conflicts in Southeast Asia ended, development and production priorities shifted back toward nuclear capable systems and their supporting research and development. However, as a result of the experience in Southeast Asia and changes in the state of the world in general, dually capable and purely conventional systems were still given some priority.

In the late 1950s, DOD established the Manufacturing Technology (ManTech) Program under the provisions of the Defense Production Act of 1950 and its extensions. The objective of this program was to strengthen the U.S. defense industrial base by encouraging the development and use of innovative manufacturing methods and processes.

COLD WAR

During the 1970s and 1980s, there was a period of relative calm in defense manufacturing. Military activity throughout the world was contained, and the pressure to develop more sophisticated weapons eased. Emphasis was placed on improving command, control, communications, and intelligence capabilities. Commercial industrial activity was focused on increasing the capabilities of computational equipment and using this equipment in manufacturing processes. The military also benefited from these advances, although the costs of manufacturing military equipment remained high. In 1975, the Secretary of Defense directed the armed services to increase their emphasis on and support for the ManTech program. Some ManTech funds were allocated to the adaptation of commercial products for use by the military in the nondevelopmental item program, the forerunner of the dual-use program.

1985 TO THE PRESENT

The disintegration of the Soviet Union and the end of the Cold War had a profound effect on the military establishment and its supporting industries. The level of expenditures that had characterized previous defense budgets was no longer politically supportable. Congress and the public at large demanded that the "peace dividend" (monies saved from defense) be used for social programs. Defense allocations plummeted and have only recently begun to level off. Large reductions of forces and significant retrenchments in research and development and procurement were necessary for the military to stay within funding constraints. Projections of greatly reduced defense budgets and advice from defense officials encouraged widespread consolidations among defense-oriented firms, with mergers and acquisitions accelerating between 1985 and 1995. The twofold objective of these consolidations was to maintain a critical mass of defense-oriented business while diversifying into the production of commercial goods and services and minimizing dependence on the defense budget.

Since 1787, the characteristics and size of the U.S. defense industrial base have changed significantly. Although, the defense industrial base certainly existed during World War II, the Korean War, and throughout most of the Cold War and the conflict in Southeast Asia, the term has become less relevant in recent years. Today, it is difficult to define exactly where commercial industry ends and the defense industrial base begins. Although many aircraft plants, arsenals, shipyards, and other industrial facilities are devoted mainly (some exclusively) to providing military hardware, many of them only assemble system components that have been manufactured elsewhere, usually by commercial industrial facilities. With the exception of munitions, the trend has been and continues to be a blurring of the line between commercial and defense industries. This trend is most apparent in the organizations that produce aircraft; space systems; command, control, and communications systems; and the infrastructure and support systems related to them.

APPENDIX B

Worldwide Web Sites and Documents Related to Defense Manufacturing

U.S. AIR FORCE

Air Force 2025
http://www.au.af.mil/au/2025/index.htm

Major Command Mission Area Plans
http://www.safmi.hq.af.mil/saf-mii/miit/pad96/pad96toc.htm

Materiel Command
http://www.afmc.wpafb.af.mil

Modernization Planning Process and Technology Master Process
http://aftech.afrl.af.mil/

New World Vistas
http://web.fie.com/fedix/vista.html

Technology Area Plans
http://aftech.afrl.af.mil/

U.S. ARMY

Strategic Technologies for the Army of the Twenty-First Century (STAR 21)
http://www.nap.edu/bookstore/enter3.cgi?mode=concept&search=STAR+21

U.S. DEPARTMENT OF DEFENSE

Assessing Defense Industrial Capabilities, Department of Defense Handbook
http://www.acq.osd.mil/iai/5000_60h/cover.htm

Defense Science and Technology Strategy
http://www.dtic.mil/dstp/DSTP/strategy/strategy.htm

Defense Sciences Office, DARPA
http://www.darpa.mil/DSO/rd/

Defense Technology Objectives
http://www.dtic.mil/dstp/DSTP/dtos/dto.htm

Dual Use Applications Program
http://www.darpa.mil/jdupo/index.html

Dual Use and Commercial Programs
http://www.acq.osd.mil/es/dut/

Electronics Technology Office, DARPA
http://www.darpa.mil/eto/RaDPrograms.html

Joint Warfighting Science and Technology Plan
http://www.dtic.mil/dstp/DSTP/jwsp.htm

Manufacturing Technology (ManTech) Program
http://mantech.iitri.com/

Military Critical Technologies List
http://www.dtic.mil/mctl/

Quadrennial Defense Review
http://www.defenselink.mil/pubs/qdr/

OTHER

Manufacturing Infrastructure: Enabling the Nation's Manufacturing Capacity
Office of Science and Technology
http://www.eng.nsf.gov/news/MireReport/mire.htm

Appendix C

Biographical Sketches of Committee Members

Alton D. Slay (chair) retired as a general from the U.S. Air Force. General Slay held a number of key positions in the Air Force, including commander, Air Force Flight Test Center, Edwards Air Force Base; director of operations requirements and development at Air Force Headquarters; deputy chief of staff for research and development; commander, Air Force Systems Command, Andrews Air Force Base; and assistant deputy chief of staff for plans and operations, U.S. Air Force in Europe. He has extensive experience in the development, deployment, and operation of weapons systems and is knowledgeable about the translation of system requirements into manufacturing capabilities.

Henry Alberts was professor of engineering management at the Defense Systems Management College and principal investigator for the college's support to the U.S. Senate Armed Services Committee on Defense Acquisition Reform issues until his retirement in June 1988. He has held positions in both industry and government related to defense technology and manufacturing and has conducted studies on acquisition process design and modification, process re-engineering, dual-use technology, and the impact of world events on the defense industrial base. His experience includes systems engineering of weapons systems, defense acquisition processes, and strategic issues in defense manufacturing. He is a certified member of the Defense Acquisition Corps and principal investigator for the U.S. Senate Armed Services Committee.

Robert F. Bescher retired as vice president, operations and manufacturing technology, at Pratt and Whitney's government engines and space propulsion unit.

Prior to that, he was vice president of manufacturing for Pratt and Whitney's operations unit, where he was responsible for the day-to-day operations of six jet engine manufacturing plants. He has extensive experience in both defense and commercial manufacturing operations. Mr. Bescher is on the Board of Directors of the National Center for Manufacturing Sciences and is the current chair of the Manufacturing Committee of the Aerospace Industries Association.

William Gibbs is manager of business systems and program manager for the Maritech Ship Project at Electric Boat Company of General Dynamics. In this capacity, he is responsible for developing manufacturing processes for a new attack submarine using state-of-the-art integrated product and process development methods. Previously, he was head of the Trident missile test program and director of integrated manufacturing systems.

Wesley L. Harris is professor of aeronautics and astronautics, director of the Lean Sustainment Initiative, and co-director of the Lean Aircraft Initiative at the Massachusetts Institute of Technology. As co-director of the Lean Aircraft Initiative, he is responsible for research to improve productivity in aircraft design and manufacturing. Previously, he was associate administrator for aeronautics at the National Aeronautics and Space Administration, where he was responsible for the strategy, planning, and direction of the aeronautics research programs. Dr. Harris is a member of the National Academy of Engineering. He has served on the U.S. Army Science Board and the National Research Council's Air Force Studies Board.

David Lando is vice president for engineering and environmental technologies at Lucent Technologies. As senior corporate manufacturing research and development executive, he is responsible for providing strategy and leadership for the corporation's manufacturing research and development technology centers of excellence. Previously, while serving as director of the Integrated Circuits Technology Laboratory and vice president for manufacturing, Asia, he was responsible for the development, deployment, and operation of worldwide integrated circuit engineering data systems. Dr. Lando's expertise includes microelectronics technology, design, and manufacturing, as well as commercial technologies and practices with the potential for improving defense manufacturing.

Aris Melissaratos is vice president of science, technology, and quality at Westinghouse Corporation, where he directs the science and technology center and the productivity and quality center. In addition, he chairs the board of the Agile Manufacturing Enterprise at Lehigh University and Science and Technology for Affordability, an industry advisory group to the U.S. Department of Defense. Previously, he held several positions in the defense electronics business at Westinghouse, including vice president and general manager of the manufacturing operations division. His expertise includes management of manufacturing operations and productivity enhancement through the application of advanced technologies and practices.

Frederick J. Michel is a consultant specializing in factory operations, factory automation, and next-generation manufacturing best practices. He served as assistant deputy for production for the U.S. Army Materiel Command and deputy chief of staff for manufacturing technology. His responsibilities included overseeing the Army's Manufacturing Technology Program. Mr. Michel is a fellow of the Society of Manufacturing Engineers and a member of the National Research Council's Board on Manufacturing and Engineering Design.

J. David Mitchell retired as vice president of manufacturing development at Rockwell International. Previously, he held the positions of vice president of strategic planning and computer-integrated manufacturing development for Rockwell Information Systems and corporate director of productivity and advanced manufacturing programs. Dr. Mitchell is chair of the Coalition for Intelligent Manufacturing Systems and a board director of Product Data Exchange Using STEP (standard for exchange of product-model data). In the past, he has served as president of the Precision Measurements Association, board director of the Robotics Institute of America, and chair of the joint Aerospace Industries Association/National Security Industrial Association Committee for Improved Army-wide Calibration Operations. His expertise includes new manufacturing processes and practices, manufacturing operations for defense products, information systems, and process automation. He is the author of more than 50 articles on advanced manufacturing technology and practices.

Deborah S. Nightingale is currently a consultant in the area of strategic planning and business development. Previously, she served as director, strategic planning and business development, for the engines division of AlliedSignal Corporation. In this capacity, she established the strategic objectives for five major business enterprises. She has also held positions in manufacturing operations at AlliedSignal and was responsible for systems technology, planning, tooling, quality, materials, and customer support. Her areas of expertise include strategic planning, manufacturing operations, information systems, computer integrated manufacturing, and computer modeling. Dr. Nightingale is a member of the National Academy of Engineering.

Dean Rhoads is a senior consultant with Arthur Anderson Consulting, where he is leading an effort to improve the software development process in terms of predictability, reliability, and productivity. Prior to joining Anderson, Mr. Rhoads held software management positions at Fidelity Investments Systems Company, Northrop Grumman Corporation, Anser Corporation, Software Productivity

Consortium, and Sperry Corporation, where he developed software for a variety of applications and improved software development processes. Mr. Rhoads also served in a number of acquisition management positions. He has taught systems engineering management in the U.S. Air Force and has published a systems engineering management guide. He is currently president and a member of the Board of Directors of the Washington Area Chapter of the International Council on Systems Engineering. He has also chaired or been a member of a number of Electronics Industry Association subcommittees for the development of software standards.

Richard Seubert is factory manager for the chemical processing facility at the Boeing Defense and Space Group, where he is responsible for all chemical processing, paint, and penetrant inspection operations. He is also the coordinator for rapid prototyping implementation strategies for all airplane programs and factory locations. Previous positions at Boeing have included responsibility for the engineering laboratories and the interface to Boeing's Digitally Driven Enterprise Initiative, which focused on single sources of product data, automated numerical control, and real-time factory floor control. Mr. Seubert is a fellow of the Massachusetts Institute of Technology Leaders for Manufacturing.